The S&L Debacle

© 1991

THE S&L DEBACLE

Public Policy Lessons for Bank and Thrift Regulation

LAWRENCE J. WHITE
NEW YORK UNIVERSITY

OXFORD UNIVERSITY PRESS
New York   Oxford

Oxford University Press

Oxford New York Toronto
Delhi Bombay Calcutta Madras Karachi
Petaling Jaya Singapore Hong Kong Tokyo
Nairobi Dar es Salaam Cape Town
Melbourne Auckland
and associated companies in
Berlin Ibadan

Copyright © 1991 by Oxford University Press, Inc.

First published in 1991 by Oxford University Press, Inc.,
200 Madison Avenue, New York, New York 10016

First issued as an Oxford University Press paperback, 1992

Oxford is a registered trademark of Oxford University Press

ISBN 0-19-506733-9
ISBN 0-19-507484-X (pbk.)

Library of Congress Cataloging-in-Publication Data
White, Lawrence J.
The S & L debacle : public policy lessons for
bank and thrift regulation /
Lawrence J. White.
p. cm. Includes bibliographical references and index.
1. Savings and loan associations—United States—Corrupt practices.
2. Savings and loan associations—Government policy—United States.
3. Savings and loan associations—United States—Deregulation.
I. Title. HG2151.W47 1991 332.3′2′0973—dc20 90-14249

2 4 6 8 10 9 7 5 3 1
Printed in the United States of America

This book is dedicated to the men and women of the Federal Home Loan Bank Board, the Federal Savings and Loan Insurance Corporation, and the Federal Home Loan Bank System, November 1986 to August 1989.

Preface

This book is about the savings and loan debacle.

From November 12, 1986, until August 18, 1989, I was one of the three Board Members of the Federal Home Loan Bank Board. The Bank Board was the regulator of and provider of deposit insurance (through the FSLIC) to the thrift industry. I had taken a leave of absence from my position as professor of economics at New York University to serve on the Bank Board; I returned to NYU at the end of my Bank Board service. This book was written after my return.

In this book I have tried to combine my skills as an economist with my experience at the Bank Board to provide a coherent analysis of the debacle: what happened, how it happened, why it happened, and what reforms are necessary so that it never happens again. The emphasis of this book is on *analysis*, rather than on personalities or anecdotes. This focus reflects my strong belief that an understanding of the economic forces and incentives that were at work is crucial to an understanding of the debacle and of the need for fundamental reform in bank and thrift regulation. I have tried to present this analysis in non-technical terms that will be comprehensible to the lay reader as well as to the specialist.

My period of service at the Bank Board was contemporaneous with most of the events and controversies included in Chapters 7, 8, and 9. These chapters cover the period when the depth of the problems of the insolvent thrifts became evident, when the Bank Board undertook major transactions to dispose of hundreds of insolvent thrifts, and when the concomitant need for substantial public financing to cover the government's insurance guarantees to depositors was fully recognized. I believe that my experience at the Bank Board helped develop and inform the insights that I have tried to convey concerning those events and controversies; in the writing of these chapters, however, I have chosen to exclude any first-person references. In the end, readers will have to decide for themselves whether my experience at the Bank Board may have colored the perspective that I provide.

I owe large debts of gratitude in many directions for assistance in making this book become a reality: to Edward Golding and Maryann Kaswell, who served as my Special Assistants at the Bank Board and who greatly increased my knowledge and understanding of the economics, politics, and institutional structure of thrift regulation and deposit insurance; to George Benston, Mitchell Berlin, Ernest Bloch, Mary Cadette, Robert Duncan, Patrick Forte, Marilyn Frankel, Devra Golbe, Edward Golding, Jay Janis, Stephen Kohlhagen, Joseph McKenzie, Jonas Prager, Herbert Sandler, Lewis Spellman, and Gregory Udell, who offered useful comments and suggestions on an early draft of the manuscript; to Ingyu Chiou, who was my research assistant and computer programmer; to Mary Jaffier and her staff, who typed the manuscript; to Suzanne Schetlick, who assisted with corrections and revisions; to Jim Cozby, whose magic produced the camera-ready copy; to Herbert Addison, whose enthusiasm for and confidence in this book made my dealings with Oxford University Press a thoroughly enjoyable experience; and especially to Judith P. Zinsser, whose constant encouragement, high standards, and tight editing were important in making this book what it is.

New York
August 1990

L. J. W.

Contents

A Brief Glossary

Most books about government make frequent references to government agencies, programs, and concepts that then become shortened to their identifying initials. This book is no exception. The following list should help readers who find themselves in the middle of a chapter and cannot remember what a particular set of initials stands for.

ARM	Adjustable rate mortgage
BIF	Bank Insurance Fund
CD	Certificate of Deposit
CEA	Council of Economic Advisers
CEBA	Competitive Equality Banking Act of 1987
DIDMCA	Depository Institutions Deregulation and Monetary Control Act of 1980
EPA	Environmental Protection Agency
FADA	Federal Asset Disposition Association
Fannie Mae	Federal National Mortgage Association
FDIC	Federal Deposit Insurance Corporation
FHA	Federal Housing Administration
FHFB	Federal Housing Finance Board
FHLB	Federal Home Loan Bank
FHLBB	Federal Home Loan Bank Board
FICO	Financing Corporation
FIRREA	Financial Institutions Reform, Recovery, and Enforcement Act of 1989
FRC	Frank Russell Company
Freddie Mac	Federal Home Loan Mortgage Corporation
FSLIC	Federal Savings and Loan Insurance Corporation
GAAP	Generally accepted accounting principles
Ginnie Mae	Government National Mortgage Association
GAO	General Accounting Office
GNP	Gross national product

HOLA	Home Owners' Loan Act of 1933
HUD	U.S. Department of Housing and Urban Development
IRA	Individual retirement account
MBS	Mortgage-backed security
MCP	Management consignment program
MMMF	Money market mutual fund
MSB	Mutual savings bank
NCUA	National Credit Union Administration
NHA	National Housing Act of 1934
NOW	Negotiable order of withdrawal
OCC	Office of the Comptroller of the Currency
OMB	Office of Management and Budget
OTS	Office of Thrift Supervision
PSA	Principal supervisory agent
QTL	Qualified thrift lender
RAP	Regulatory accounting principles
REFCORP	Resolution Funding Corporation
RTC	Resolution Trust Corporation
SAIF	Savings Association Insurance Fund
Sallie Mae	Student Loan Marketing Association
S&L	Savings and loan association
VA	Veterans Administration

The S&L Debacle

CHAPTER ONE

Introduction

The S&L debacle—the massive insolvencies of hundreds of savings and loan associations—has been a traumatic event for the worlds of finance and government in the United States. The costs of dealing with this mess—of honoring the government's deposit insurance obligations—will be huge and seem to be ever rising. Revelations of new financial abuses or political intrigues related to the debacle are almost a daily staple of the media.

This book is about that debacle.[1] It offers a perspective on these large-scale insolvencies and their huge costs: what happened, how it happened, why it happened—and what must be done to ensure that it never happens again. It is a book about fundamental economic forces and incentives and about government policies that went awry and that, in different ways, continue to go awry. The public policy lessons that need to be learned from this debacle are broad and go far beyond regulation and deposit insurance for the thrift industry alone. They apply with equal validity to the regulation of and deposit insurance for the much larger world of commercial banking. They also have value for the reform of bank regulation in other countries, who can learn from the sorry and extremely costly mistakes of the U.S. government.

The fields of bank and thrift regulation and the operations of their deposit insurance systems are arcane and are usually of little interest to anyone outside of those areas. An understanding of these processes, however, is crucial to comprehending what went wrong, why the thrift deposit insurance arrangement went so badly astray, and why a major overhaul of our regulation and deposit insurance systems is so vital. This book will attempt to provide that understanding.

A number of themes will surface throughout this book. First, the bank and thrift regulatory (and deposit insurance) regimes of the United States are based on a fundamentally flawed information system. That information system—the standard accounting framework used by bank and thrifts—looks backward at historical costs rather than at current market values. It does not yield the information that would allow regulators to protect the deposit insurance funds properly. Among the serious policy mistakes of the early 1980s were the decisions to weaken this already inadequate information system for thrifts. The revamping of this accounting framework—a switch to market value accounting—is the single most important policy reform that must be accomplished. All else pales in comparison.

Second, the unwillingness of the Congress and the regulators to treat deposit insurance as *insurance* has been unfortunate and extremely costly. The principles and protections of insurance are valuable and have direct application to deposit insurance. The policy actions of the early 1980s to weaken the system of safety-and-soundness regulation for thrifts—the rules and enforcement system that protects the deposit insurance fund—can only be described as perverse. Even today, the reluctance of government policymakers even to consider replacing the current flat-rate deposit insurance premiums with a system of risk-based premiums—a standard practice and protection in virtually all other areas of insurance—is an indication of a continuing mindset.

Third, the economic deregulation that the federal government and many states provided for the thrift industry in the early 1980s was basically sound. But, to make that deregulation work properly, the federal government needed to strengthen the safety-and-soundness regulatory regime, improve the information system, and focus more on economic incentives. Unfortunately, exactly the opposite occurred in all three areas. In the absence of tighter scrutiny and the appropriate incentives and controls, hundreds of thrifts took advantage of the new opportunities and expanded rapidly with new loans and investments; in all too many instances these thrifts' executives were overly aggressive, excessively optimistic, careless, ignorant, and/or fraudulent and criminal. The decline in the price of oil and changes in the tax laws affecting real estate in the 1980s compounded what was already going to be an extremely difficult situation.

Fourth, the regulatory system was tightened (albeit imperfectly) and the rampage was largely halted by 1986. But the costs of the shoddy loans and investments—the writedowns of their values, and their inability to yield adequate incomes to cover the interest on the insured deposits that funded them—have been recognized only slowly since then. The revelation of these costs are often misinterpreted to mean that abuses are still occurring rather than that the accounting system is slow to register these embedded costs.

Fifth, the necessity to spend large sums to clean up the problems of the insolvent thrifts has been widely misunderstood and mischaracterized. The term *bailout* is almost always used to describe this cleanup. It is a wholly inappropriate term that conveys the wrong connotations and implications. The moneys are being used to satisfy the government's insurance obligations to depositors, not to "bail out" anyone. The term has stuck, however, and continues to shape thoughts, policies, and actions in inappropriate ways.

Sixth, the cleanup legislation passed in August 1989—the Financial Institutions Reform, Recovery, and Enforcement Act—did not provide adequate funds for the cleanup and did not address most of the necessary fundamental reforms of bank and thrift regulation and deposit insurance. And, along the way, the legislation created substantial confusion and disruption, worsened the financial condition of the healthy majority of the thrift industry, and added to the costs of the cleanup.

The Congress and the Bush administration will have to readdress these issues of funding, of the fundamental reform of bank and thrift regulation and deposit insurance, and of the financial future of the thrift industry—probably in 1991. The lessons of the debacle need to be learned; they form the basis for the fundamental changes that are crucial for the 1990s and beyond.

An Overview of the Book

Chapter 2 provides a brief description of the thrift industry as it stands today and a discussion of recent trends. The thrift industry has been and continues to be a major part of the financial services sector of the U.S. economy. Thrifts continue to be major providers of deposit services and of mortgage originations and financing for households. The massive insolvencies of hundreds of thrifts have colored the recent aggregate financial figures of the thrift industry and have masked the fact that a substantial segment of the industry is solvent and profitable, although some of them are only thinly so.

Chapter 3 describes the deposit insurance and regulatory structure that applies to banks and thrifts. The framework is extensive and intricate. The major information flow supporting that regulatory framework is the accounting information generated by those banks and thrifts. Because that information is so crucial for understanding the problems of bank and thrift regulation, the chapter offers a brief primer on bank and thrift accounting. A number of extremely important lessons and insights for the problems of regulation follow from an understanding of this simple accounting example.

Chapter 4 recounts the financial and political history of the thrift industry through the late 1970s. The industry was an important part of the residential mortgage finance system of the U.S. It was sleepy, prosperous, and safe. But it had one serious flaw: Thrifts were taking in short term-deposits and lending these funds on long-term mortgages, which made them quite vulnerable to sharp increases in interest rates. When the problem first arose in the mid-1960s, the Congress applied a regulatory patch: ceilings on deposit interest rates. This patch worked for more than a decade. When interest rate increases returned with a vengeance in the late 1970s, however, the deposit rate ceilings could no longer be effective, and the thrifts' hemorrhaging began.

Chapter 5 describes the interest rate squeeze of the late 1970s and early 1980s. It then outlines the policy responses at both the federal and state levels—*economic* deregulation, which widened the economic powers and activities that were open to thrifts. These economic deregulatory steps were largely sensible, but they needed to be accompanied by strengthened *safety-and-soundness* regulation. Unfortunately, virtually no attention was given at the time to the need for strengthening this latter type of regulation. Indeed, a number of perverse federal policy actions actually weakened the existing system of safety-and-soundness regulation. These regulatory actions and inactions, on top of the financial stresses that had already enveloped many thrifts, created a set of opportunities, capabilities, and incentives for risk-taking that proved irresistible (and eventually disastrous) for hundreds of thrifts.

Chapter 6 explores the behavior that followed. Many thrifts, especially state-chartered thrifts in the sunbelt (stretching from Florida through Louisiana and Texas to California), grew rapidly in the 1983-1985 period. Rapid growth meant a rapid expansion of assets and of deposit liabilities, which were insured by the Federal Savings and Loan Insurance Corporation (FSLIC). If the new assets (loans and investments) held their value and yielded profits, all was well. If they fell in value, the FSLIC was obligated to cover the shortfall to the depositors. Unfortunately, the latter occurred in hundreds of thrifts. The causes were an amalgam of deliberate risk-taking, carelessness, poor management, excessive optimism, bad luck, and fraud and criminal activity, compounded by the decline in the price of oil in the mid-1980s (and its consequences for the economies of energy-oriented states, especially Texas) and by changes in the tax laws.

Chapter 7 reviews the regulatory efforts of the Federal Home Loan Bank Board to contain the damage. After a false start in 1984 that tried (misguidedly) to reduce rapidly growing thrifts' reliance on deposit brokers as a source of supply of the deposits that were funding that growth, the Bank Board turned in 1985 through 1988 to regulations that limited directly the asset growth and risk-taking that were occurring. Further, it increased sharply the number of examiners and supervisors—its field-force regulatory staff. Insolvent thrifts that could not be closed (because of inadequate FSLIC funds to cover the accumulated losses) were stabilized through tight supervisory controls or (where necessary) through the replacement of their owners and executives with Bank Board-appointed managers. These improvements, reinforced by a generally heightened regulatory concern for safety and soundness, largely reined in the excessive risk-taking. These regulatory changes, however, could not cure or reverse the costly investment mistakes that too many thrifts had already made. The delays in regulatory tightening were paralleled by delays in recognition, first by the Bank Board and then by the Congress, that the FSLIC was headed for insolvency. Thrift industry lobbying helped lengthen this delay. It was August 1987 before the Congress finally passed legislation that authorized the FSLIC to "recapitalize" itself by borrowing $10.8 billion against future insurance premium revenues—but that also limited the FSLIC to borrowing no more than $3.75 billion in any twelve-month period and called for forbearance in closing some insolvent thrifts.

Chapter 8 analyzes the activity of the Bank Board in disposing of insolvent thrifts, especially during 1988. The FSLIC liquidated or found acquirers for more than 200 thrifts during 1988, in a set of transactions that generated (and continues to generate) considerable controversy. The disposals explicitly obligated the FSLIC to future payments of tens of billions of dollars to honor the insurance guarantees that had been made to depositors in these thrifts. Criticisms of these transactions often failed to acknowledge that the FSLIC was already implicitly obligated to cover the costs of honoring these guarantees; the transactions simply made the costs explicit. Tax breaks for acquirers were also involved, which made these transactions yet more controversial. An analysis of the structure of the transactions and of the available evidence indicates that in aggregate they were cost-effective and financially sound.

Chapter 9 discusses the new cleanup legislation, which was proposed by President Bush in February 1989 and enacted by the Congress in August. The legislation authorized an additional $50 billion in borrowing in order to fund the disposals of the remaining insolvent thrifts. The healthy part of the thrift industry was expected to cover a significant fraction of the overall costs of the cleanup, through the payment of higher insurance premiums and the taxation of a portion of the capital and future earnings of the Federal Home Loan Banks. The bulk of the costs, however, would be added to the burden on general tax revenues. The Bank Board and the FSLIC were abolished, and new agencies were created. The legislation specified a set of net worth standards for thrifts and numerous changes in the details of the regulations that apply to thrifts' operations. It allocated extra moneys for a stepped-up effort to bring criminal prosecutions of errant thrift executives. In all, the legislation ran to 381 printed pages. Unfortunately, it also created at least as many problems as it solved, and it left most of the important regulatory reforms largely unaddressed.

Chapter 10 steps back from the story of the S&L debacle and asks some fundamental questions about deposit insurance and the bank and thrift regulation that accompanies it. Why have deposit insurance? Should there be limits on it? What kind of regulation should accompany it? Are there lessons for deposit insurance that can be learned from the practices and procedures that apply to other forms of insurance? The chapter demonstrates that deposit insurance does have a proper role to play and that there are valuable lessons that can be learned from other forms of insurance.

Chapter 11 addresses the fundamental reforms that are necessary for bank and thrift regulation and deposit insurance. Foremost is a change in the information system used for regulation: Market value accounting must replace the existing, historical-cost-based accounting system. This improved information system would greatly enhance the effectiveness of the higher net worth standards that are required for banks and thrifts in the 1990s. Second, the risk-based elements of those net worth standards must be improved. Third, risk-based premiums must replace the current system of flat-rate premiums, which ignore risk and thereby provide no incentive for banks and thrifts to refrain from exploiting risk-taking opportunities. Incentives matter. Risk-based premiums can help discourage risk-taking and encourage banks and thrifts to maintain net worth levels above the mandated minimums. Fourth, the powers of the regulators to intervene and take control of an errant bank or thrift, before it reaches insolvency, must be clarified and strengthened. Fifth, deposit insurance should not be cut back, but should instead be expanded to cover all deposits. This expansion is the only way to deal with the problem of bank runs. In contrast, any reductions in coverage would exacerbate the problem of runs. Other reforms include the expanded use of long-term subordinated debt; the designation of the U.S. Treasury as the automatic backup for the deposit insurance funds; a clarification of which activities should be allowed within an insured bank or thrift and which should be permitted for holding companies; the expanded use of data and evidence to support the regulations; and expanded education of bank and thrift executives.

Finally, Chapter 12 explores two themes that are directly relevant to the necessary reforms: the rapidly changing and improving technologies that undergird banks and thrifts, which will surely lead to more competition within and among them; and the future role of thrifts in the financial services markets. On this latter topic, the chapter argues that the headlines of the debacle overlook the group of healthy, efficient thrifts that can compete effectively in the marketplace. These thrifts should be neither protected and coddled nor taxed and regulated to extinction. Rather, they should be allowed the flexibility to adapt their skills to the changing marketplace. The chapter and book conclude by reviewing the important policy lessons that can, and must, be learned from the S&L debacle.

Notes

1. Other recent books on the thrift debacle include Brumbaugh (1988), Strunk and Case (1988), Eichler (1989), Kane (1989), Pizzo, Fricker, and Muolo (1989), Pilzer (1989), and Adams (1990). See also Brumbaugh and Carron (1987), Brumbaugh and Litan (1989), and Brumbaugh, Carron, and Litan (1989).

PART ONE

Background

CHAPTER TWO

The Thrift Industry: A Recent Picture

Despite the severe losses experienced by hundreds of thrifts, the thrift industry has been and remains an important part of the financial services sector of the U.S. economy. The industry accounts for a significant share of the assets of all depository institutions and of deposits held by the public in those institutions. It continues to be the industry group that accounts for the largest share of current mortgage originations, and thrifts collectively are the holders of the largest share of the existing stock of mortgages.

One distinction within the thrift industry should be noted at the beginning. Prior to 1980 the thrift industry could be readily divided into two discernible segments: savings and loan associations (S&Ls) and savings banks. S&Ls accounted for the largest part of the industry. Most of them were insured by the Federal Savings and Loan Insurance Corporation (FSLIC), though some were insured instead by state-sponsored private insurance arrangements. Their investment activity was focused almost exclusively on mortgage lending; more than 75 percent of their asset portfolios were devoted to residential mortgages. They were split about evenly between state-chartered and federally chartered associations. Though most had a mutual form of ownership (i.e., they were nominally owned by their depositors), some (especially the larger California institutions) were stock associations (i.e., they were owned by stockholder investors). The other segment of the thrift industry was composed of savings banks, most of which were insured by the Federal Deposit Insurance Corporation (FDIC), which also insured commercial banks. Savings banks were located largely in the Northeast (and the state of Washington) and were exclusively state-chartered and mutual in organization (and hence were referred to as mutual savings banks, or MSBs). Though also heavily involved in mortgage lending, they had been involved traditionally in other forms of consumer lending and securities investments as well.

Most of these distinctions blurred during the 1980s. FSLIC-insured thrifts could acquire savings bank state charters or, in some instances, just change their names to call themselves a "savings bank." Some of the FDIC-insured savings banks acquired federal charters and/or converted to stock associations. With the deregulation of the early 1980s FSLIC-insured thrifts were allowed to diversify somewhat away from residential lending.

The only major distinction between the two groups that has remained has been the identity of the insurance fund. Prior to August 1989 the larger group was insured by the FSLIC; the smaller group was insured by the FDIC. Since August 1989 both groups have been insured by the FDIC, but their insurance funds within the FDIC are still separated: The former group is part of the Savings Association Insurance Fund (SAIF), while the latter group (along with commercial banks) are part of the Bank Insurance Fund (BIF).[1]

Because the widespread insolvencies and huge losses were almost exclusively from the group of FSLIC-insured thrifts,[2] most of this book will focus on the group of formerly FSLIC-insured and now SAIF-insured insured thrifts (and these will sometimes be referred to as S&Ls, though some technically are savings banks). "The thrift industry," however, broadly encompasses the BIF-insured MSBs as well, and the data concerning the industry in this chapter will sometimes refer to both groups.

The Relative Size and Importance of the Thrift Industry

The thrift industry is a major provider of two important financial services or "products." Thrifts are places where households keep deposits and where they borrow money for residential mortgages. Many thrifts, of course, provide other services, such as credit cards, car loans, safe deposit boxes, and insurance. But deposits and mortgage loans are thrifts' two primary services.

Table 2-1 shows the level of thrifts' deposits as of the end of 1989 and their relative importance among all depository institutions.[3] As can be seen, SAIF-insured thrifts had slightly less than $1 trillion in deposits and accounted for slightly less than a quarter of all deposits in depositories. When measured by their aggregate assets—a more common indicator of a financial institution's size—these thrifts to-taled over $1.25 trillion in assets and accounted for slightly more than a quarter of the total for all depositories. More than 280 of these SAIF-insured thrifts were insolvent at the end of 1989, however, and a few hundred more were likely to sink into insolvency in 1990 and 1991. The disposals of these insolvents, and the necessity of many of the remaining solvent thrifts to shrink, so as to raise their net worth percentages, will surely mean overall shrinkage for the aggregate group of SAIF-insured thrifts in the early 1990s.

There are two ways of representing a financial institution's activity and importance with respect to mortgages. One way is to indicate the volume of originations, or new mortgage loans made, during the year; the other is to indicate the total value of the mortgages held by the institution as of the end of the year. The former represents services for new borrowers; the latter represents loan financing services for all existing borrowers.

Tables 2-2 and 2-3 provide recent data for both types of mortgage services, for both single-family[4] and multifamily residential mortgages. As can be seen, SAIF-insured thrifts were the largest group of originators and holders of residential mortgages. They accounted for somewhat less than half of all single-family mortgage originations in

Table 2-1
Number, Deposit Liabilities, and Assets of Depository
Institutions, 1989[a]

	Number	Deposit Liabilities		Assets	
		Amount (billions)	Share of Total	Amount (billions)	Share of Total
Commercial banks	12,706	$2,548	66.1%	$3,299	66.2%
Thrifts:					
SAIF-insured[b]	2,878	946	24.5	1,252	25.2
BIF-insured	469	195	5.1	241	4.8
Credit unions[c]	13,371	167	4.3	184	3.7
Total	29,424	$3,856	100.0%	$4,976	100.0%

[a] As of December 31.
[b] Formerly FSLIC-insured.
[c] Federally insured.

Sources: FDIC, OTS, and NCUA data.

1989 and for more than a third of the holdings of this category of mortgage. Commercial banks were the next largest group in both categories. Mortgage bankers, who specialize in originating mortgages, but who then sell virtually all of those mortgages to others, were the third largest group of originators. A similar story was true for multifamily residential mortgages.

One further point concerning the mortgage origination data should be made, however. Many thrifts originate mortgages through subsidiaries or affiliates that are separately organized from the thrift (this is true for many commercial banks as well); these originations are included in the mortgage banking category rather than in the thrift category. Unfortunately, precise data are not available, but it has been estimated that inclusion of their subsidiary and affiliate originations in the thrift totals would bring the thrift share of aggregate originations above 50 percent.[5]

Table 2-2
Single-Family[a] Residential Mortgage Originations,
1989, and Holdings, 1988

	Originations[b]		Holdings[c]	
	Amount (billions)	Share of Total	Amount (billions)	Share of Total
Thrifts:				
SAIF-insured	$135	38%	$779	36%
BIF-insured	23	7	108	5
Commercial banks	123	35	410	19
Mortgage bankers	66	19	n.a.	n.a.
Others	5	1	864	40
Total	$352	100%	$2,161	100%

[a] One-to-four family dwellings.
[b] During 1989.
[c] As of December 31, 1988; includes holdings of mortgage-backed securities and construction loans.

Sources: Barth and Freund (1989); HUD data.

Important Characteristics of
SAIF-Insured Thrifts, Circa 1989

At the beginning, it is useful to realize that the SAIF-insured thrifts in 1989 were essentially two dichotomous groups: a majority of profitable and solvent thrifts and a sizable minority of unprofitable thrifts, many of which were insolvent or soon-to-be insolvent. Table 2-4 shows the two groups and their important characteristics. The profitable group of 1,812 thrifts constituted about 60 percent of the total, with $753 billion in assets. They were solvent, and their profits constituted a respectable return on their assets and on net worth, especially considering the difficult operating environment that they faced in 1989. The remainder were unprofitable; most of them were insolvent or would soon be so. Their losses during 1989 were huge and greatly overwhelmed the profits of the majority. Two-thirds of these losses were nonoperating losses: capital losses on asset sales and writedowns of assets, both of which were the recognition of the loss of value of loans and investments made in earlier years. The remaining one-third were operating losses, largely attributable to the inadequate incomes generated by these same loans and investments and also to the

Table 2-3
Multifamily[a] Residential Mortgage Originations and
Holdings, 1988

	Originations[b]		Holdings[c]	
	Amount (billions)	Share of Total	Amount (billions)	Share of Total
Thrifts:				
FSLIC-insured	$18	46%	$88	30%
FDIC-insured	3	8	19	7
Commercial banks	7	18	37	13
Mortgage bankers	5	12	n.a.	n.a.
Other	6	16	150	50
Total	$38	100%	$293	100%

[a] Five-or-greater family dwellings.
[b] During 1989.
[c] As of December 31, 1988; includes mortgage-backed securities.

Sources: Barth and Freund (1989); HUD data.

somewhat higher interest rates that these thrifts had to pay to depositors who were nervous about the safety of their deposits, despite the presence of government deposit insurance.

The average size of the thrifts in the overall group masks a substantial dispersion of sizes. Approximately half the group were less than $100 million in size (by assets), and well over a quarter of the remainder were between $100 million and $400 million in size. The tradition of many small thrifts, most of which are mutual in organization (i.e., "owned" by their depositors) is an important characteristic of the industry and one that will be referred to again in later chapters.

Recent Trends

The preceding pages have provided a 1989 "snapshot" of the thrift industry, especially the SAIF-insured thrifts. Some recent trends for the latter group will add perspective to this picture.

First, the industry's aggregate financial fortunes have fluctuated sharply since the mid-1970s. Table 2-5 provides annual data on net income (i.e., annual profit or loss) as a percentage of assets and tangible net worth[6] as a percentage of assets. As can be seen both

Table 2-4
Selected Characteristics of SAIF-Insured Thrifts, 1989

	Profitable	Unprofitable	Total
Number	1,812	1,066	2,878
Aggregate Assets (billions)	$753	$499	$1,252
Average size (billions)	$416	$468	$435
Deposits (billions)	$553	$393	$946
As a percentage of total assets	73.5%	78.8%	75.6%
Net worth (capital) (billions)[a]	$33.8	$-23.7	$10.1
As a percentage of total assets	4.6%	-4.7%	0.8%
Net income after taxes (billions)	$5.2	$-24.4	$-19.2
As a percentage of total assets	0.7%	-4.9%	-1.5%

[a]Tangible net worth; excludes goodwill as an asset.

Source: OTS data.

measures were healthy through 1979, deteriorated sharply beginning in 1980, revived in the mid-1980s, and plunged again in the late 1980s. (The increase in net worth in 1988 was due to the FSLIC's disposals of 205 insolvent thrifts and the stabilization of another 18; also, acquirers of 179 of these insolvents brought $3 billion of new capital to these institutions.) Unlike the experience of the early 1980s, which was one of widespread losses, the losses of the late 1980s were by a minority of the industry; however, those losses were so large that they swamped the profits of the rest of the industry. This different pattern in the late 1980s can be conveyed in two ways. In Table 2-6 the first two columns show the percentages of FSLIC-insured thrifts that were profitable in each year between 1978 and 1989. As can be seen, in 1981 fewer than one-sixth of all thrifts were profitable and in 1982

Table 2-5
Net Income and Tangible Net Worth, as a Percentage
of Assets, for FSLIC-Insured Thrifts, 1975–1989

	Net Income as a Percentage of Assets	Tangible Net Worth as a Percentage of Assets
1975	0.4%	5.8%
1976	0.6	5.6
1977	0.7	5.5
1978	0.8	5.6
1979	0.7	5.7
1980	0.1	5.3
1981	−0.7	4.0
1982	−0.6	0.5
1983	0.2	0.4
1984	0.1	0.3
1985	0.3	0.8
1986	0.01	1.3
1987	−0.6	0.7
1988	−0.9	1.7
1989[a]	−1.5	0.8

[a] SAIF-insured thrifts.

Sources: FHLBB and OTS data.

fewer than one-third were in that category. By contrast, in 1989 more than three-fifths of all thrifts were profitable, though some were only thinly so. The remaining columns show the relevant percentages for the group within the industry that were solvent on the basis of their tangible net worth. This group declined through 1984, but has stabilized since then. Many of these tangible-solvent thrifts are only barely solvent and are likely to disappear in the next few years. Nevertheless, as this table and Table 2-4 show, the losses and deep insolvency of the minority of thrifts are far from representative of the entire industry.

A second important trend is the involvement of the thrift industry in residential mortgage finance, especially since many observers believe that this is the industry's main reason for existing. As noted earlier, thrifts' activity in mortgage markets can usefully be measured

Table 2-6
Profitable FSLIC-Insured Thrifts and Tangible-Solvent
Thrifts, 1978–1989

	Profitable Thrifts		Tangible-Solvent Thrifts		
	Number as a Share of All Thrifts	Assets as a Share of All Thrift Assets	Number as a Share of All Thrifts	Assets as a Share of All Thrift Assets	Tangible Net Worth as a Percentage of Assets
1978	97.3%	98.7%	99.1%	99.9%	5.6%
1979	93.5	95.9	99.2	98.9	5.7
1980	64.4	67.0	98.9	99.9	5.3
1981	15.2	8.7	97.1	96.1	4.2
1982	32.2	39.4	87.3	67.9	3.5
1983	64.8	66.8	83.6	65.0	3.4
1984	72.1	73.3	78.1	63.3	3.2
1985	78.8	84.9	78.6	67.1	3.6
1986	73.0	72.4	79.3	70.5	4.2
1987	64.8	66.2	78.6	71.7	4.3
1988	69.7	68.2	83.0	78.5	4.1
1989[a]	63.0	60.1	82.0	76.7	4.6

[a] SAIF-insured thrifts.

Sources: FHLBB and OTS data.

both by their share of origination activity and by their share of mortgages held. Table 2-7 presents the annual figures for 1970 and later. The first column shows an apparent peaking of thrifts' share of originations in 1976 and a substantial decline since then. As noted earlier, however, a significant fraction of thrifts' originations occur through subsidiaries and affiliates and are not recorded in these data; it is likely that this fraction has increased in the past decade. Thus, though the thrifts' (plus subsidiaries and affiliates) share of originations probably has declined, the decrease in share is probably less

Table 2-7
FSLIC-Insured Thrifts' Involvement in Home
Mortgage Loans, 1970–1989

	Thrifts' Originations as a Share of All Originations	Thrifts' Holdings as a Share of All Holdings	Thrifts' Holdings as a Share of Thrifts' Total Assets
1970	42%	41%	71%
1971	46	—	—
1972	48	—	—
1973	49	46	71
1974	46	—	—
1975	53	—	—
1976	55	49	69
1977	53	—	—
1978	49	—	—
1979	44	47	71
1980	46	46	71
1981	43	44	70
1982	36	42	64
1983	40	42	62
1984	48	43	59
1985	44	40	56
1986	40	37	55
1987	39	36	57
1988	43	36	58
1989[a]	38	n.a.	n.a.

[a] SAIF-insured thrifts.

Sources: Barth and Freund (1989); HUD data.

than is indicated in the table. The second column of Table 2-7 shows a substantial decline since the mid-1970s in the fraction of all home mortgages that are held by thrifts (including their holdings of mortgage-backed securities). The extent of the decline depends on the years chosen for comparison, but the trend is clear.

This decline could have occurred because thrifts grew more slowly than did the mortgage markets or because thrifts devoted a lesser share of their resources to mortgages. It appears that both forces were at work. Between 1976 (the high point of FSLIC-insured thrifts' share of holding home mortgages) and 1988, the total of home mortgage holdings grew at an annual compound rate of 12.3 percent; the aggregate of FSLIC-insured thrifts' assets grew at an annual rate of only 11.1 percent during that same span of years. Thus, thrifts grew more slowly than did aggregate mortgage holdings. As the last column of Table 2-7 reveals, however, the fraction of thrifts' aggregate assets devoted to home mortgages fell sharply after 1980. If thrifts had maintained the same fraction of their assets in home mortgages in

Table 2-8
Trends in Mutual Organization and in State Charters,
FSLIC-Insured Thrifts, 1960–1989

	Mutual Organization		State-Chartered	
	Number as a Share of All Thrifts	Assets as a Share of All Thrifts' Assets	Number as a Share of All Thrifts	Assets as a Share of All Thrifts' Assets
1960	87%	85%	54%	43%
1965	85	78	55	46
1970	85	79	53	44
1975	85	79	50	41
1980	80	73	50	44
1985	67	44	47	36
1986	63	38	46	36
1987	60	30	44	35
1988	56	26	42	29
1989[a]	54	24	42	26

[a] SAIF-insured thrifts that were not in FDIC receiverships or conservatorships.

Sources: FHLBB and OTS data.

1988 as they had held in the mid-1970s, their share of home mortgage holdings in 1988 would have been 44 percent rather than the actual level of 36 percent.

This decline in the relative importance of thrifts in the mortgage market (and of mortgages for thrifts) was not an accidental or random occurrence. As will be seen in Chapter 5, the early 1980s were a period of federal and state legislative and regulatory actions that were specifically designed to allow thrifts to diversify their asset portfolios away from (and thus reduce their dependence on) the narrow specialty of residential mortgage finance. It should come as no surprise that thrifts did what public policy expected them to do. (The poor choices of alternative investments that were made by hundreds of thrifts, however, were not expected by public policy.)

Despite these declining trends, the 1988-1989 snapshot presented earlier shows that the thrift industry still accounts for the largest percentage of mortgage originations and holdings among the identifiable groups of finance industry participants.

Two other trends are worth noting. First, the industry has been undergoing a process of transformation away from the mutual form of organization and toward stock companies. This pattern is shown in Table 2-8. The pace of conversions clearly accelerated after 1980.[7] The second trend, also shown in Table 2-8, is toward federal charters and away from state charters. Thrifts have usually had a great deal of flexibility in switching between state and federal charters, so they frequently chose on the basis of perceived advantages for their intended business strategies. Since the recognition in the mid-1980s that state-chartered thrifts were disproportionately represented among the insolvents, however, the federal regulatory apparatus has leaned heavily toward federal charters for thrifts that had to be placed under tight supervisory controls and for the new charters that were being issued to the acquirers of the insolvents. Also, the Financial Institutions Reform, Recovery, and Enforcement Act of 1989 greatly reduced the advantages that could accrue to being a state-chartered thrift, so switches to federal charters are likely to accelerate in the early 1990s.

Notes

1. The two separate funds (SAIF and BIF) were created by the FIRREA legislation in order to justify the continued levying of higher insurance

premiums on the thrifts that were formerly insured by the FSLIC; see Chapter 9.

2. The MSBs were state-chartered thrifts in states where bank regulatory officials had a more conservative approach to banking. Thus they "missed out" on the debacle that followed. In 1989, however, more than one-quarter of the MSBs ran losses aggregating $1.3 billion and causing the overall MSB group to register a loss of $97 million.

3. Money market mutual funds (MMMFs), which had approximately $400 billion in shares (which are quite similar to deposits) at the end of 1989, have been excluded from this table for a number of reasons. First, their shares are not government insured. Second, they do not offer the same type of "bricks and mortar" convenience that the depositories included in the table offer to most of their depositors. Third, a significant fraction of MMMF funds are re-invested in bank and thrift certificates of deposit, so a substantial degree of double-counting would be entailed.

4. Technically, "single-family" or "home" mortgages encompass loans on structures that accommodate one, two, three, or four families. "Multi-family" encompasses loans on structures with five or more families.

5. Barth and Freund (1989, p. 83).

6. Net worth, when measured according to generally accepted accounting principles (GAAP), allows the inclusion of "goodwill" as an asset. Goodwill can represent going-concern or "franchise" value and thus represent value that could be salvaged if the thrift were sold; in other instances goodwill may be pure artifact that has no salvage value. Tangible net worth excludes all goodwill as an asset. A more fundamental flaw in GAAP accounting, even with the exclusion of goodwill, is that it is historical cost-based and generally does not try to capture the current market values of assets, liabilities, and "off balance sheet" items.

7. The numbers and assets percentages for mutuals for 1985 and after were raised slightly by the Bank Board's management consignment program (MCP). The MCP was used to stabilize insolvent thrifts, the managements of which were thought to be highly unreliable and unlikely to be restrained even by tight supervisory controls. A "pass-through" receivership would be arranged, and the thrift would continue to operate under a federal mutual charter, with new, trustworthy management that was usually "consigned" from other thrifts; see Chapter 7. At the end of 1988, the MCP mutual thrifts (plus the 1988 "stabilizations" that were never formally called MCP thrifts) constituted 1 percent of all thrifts and 1 percent of all thrift assets.

CHAPTER THREE

The Regulatory Framework and the Importance of Accounting

Depository institutions are among the most heavily regulated entities in the American economy. Banks, thrifts, and credit unions are generally restricted as to where they can do business, what kind of business they (or their subsidiaries, affiliates, or holding companies) can do, and how they can do it. The regulatory framework encompasses a complex interleaving of federal and state laws, regulations, and regulatory bodies.

Any effort to show graphically these complex relationships among regulatory agencies and the depositories under their jurisdiction runs the risk of drawing a diagram that more resembles Piet Mondrian's painting "Broadway Boogie Woogie" than the neat organization charts shown in civics textbooks.[1] When the Federal Reserve Bank of New York tried to describe these relationships in 1988, it chose a spreadsheet format, which is reproduced in Table 3-1. The complexity of the spreadsheet speaks for itself.

This extensive regulation is a reflection of the American political system's long-standing and deeply rooted belief that depository institutions are "special": that their crucial roles as providers of credit (loans) to businesses and households and as repositories for deposit moneys by the rest of the economy make them special. One important strand of American populism has embodied a deep distrust of financial institutions, especially banks, which has led to extensive regulatory limitations on their activities. Further, their role as issuers of deposits has led to a public concern that these institutions be safe places for individuals and businesses to leave their money. In turn, this latter concern has motivated not only extensive regulation, but also a history of government-sponsored or government-operated insurance funds for the deposits in these institutions. The history of state involvement in deposit insurance extends back

Table 3-1
Depository Institutions and Their Regulators

	Chartering and Licensing 1.	Branching		Mergers, Acquisitions and Consolidations		Reserve Requirements 6.
		2. Intra-state	3. Inter-state	4. Intra-state	5. Inter-state	
A. National Banks	Comptroller	Comptroller	(4)	Comptroller (8)	(16)	Federal Reserve (18)
B. State Member Banks	State authority	Federal Reserve & state authority	(4)	Federal Reserve & state authority (9)	(16)	Federal Reserve (18)
C. Insured State Nonmember Banks	State authority	FDIC & state authority	(5)	FDIC & state authority (10)	(16)	Federal Reserve (18)
D. Noninsured State Banks	State authority	State authority	(5)	State authority (11)	State authority	Federal Reserve (18)
E. Savings Banks Federal Mutual	FHLBB	FHLBB	FHLBB (6)	FHLBB	(16)	Federal Reserve (18)
Savings Banks State Mutual	State authority	FDIC & state authority	FDIC & state authority	FDIC & state authority (12)	(16)	Federal Reserve (18)
F. Savings & Loan Associations—Federal	FHLBB	FHLBB	FHLBB (6)	FHLBB	(16)	Federal Reserve (18)
Savings & Loan Associations—State	State authority	FHLBB & state authority	FHLBB & state authority	FHLBB & state authority	(16)	Federal Reserve (18)
G. Credit Unions Federal	NCUAB	(2)	(2)	NCUAB	(16)	Federal Reserve (18)
Credit Unions State	State authority	State authority	State authority	NCUAB & state authority (13)	(16)	Federal Reserve (18)

Table 3-1 (continued)
Depository Institutions and Their Regulators

	Chartering and Licensing 1.	Branching 2. Intra-state	3. Inter-state	Mergers, Acquisitions and Consolidations 4. Intra-state	5. Inter-state	Reserve Requirements 6.
H. Bank Holding Companies	Federal Reserve & state authority	Federal Reserve & state authority	Federal Reserve & state authority	Federal Reserve & state authority	Federal Reserve & state authority (17)	N/A
I. Foreign Branches of U.S. Banks National & State Members	Federal Reserve & state authority	N/A	N/A	N/A	N/A	(19)
Foreign Branches of U.S. Banks Insured State Nonmembers	FDIC & state authority	N/A	N/A	N/A	N/A	N/A
J. Edge Act Corporation	Federal Reserve	Federal Reserve	Federal Reserve	Federal Reserve (14)	Federal Reserve (14)	Federal Reserve (18)
Agreement Corporations	State authority (1)	Federal Reserve	Federal Reserve	Federal Reserve (14)	Federal Reserve (14)	Federal Reserve (18)
K. International Banking Facilities	N/A	N/A	N/A	N/A	N/A	N/A
L. U.S. Branches & Agencies of Foreign Banks—Federal	Comptroller	Comptroller & FDIC (3)	Comptroller & Federal Reserve (7)	Comptroller & Federal Reserve (15)	(15)	Federal Reserve (18)
U.S. Branches & Agencies of Foreign Banks—State	State authority	State authority & FDIC (3)	State authority & Federal Reserve (7)	FDIC Federal Reserve or state authority (15)	(15)	Federal Reserve (18)

Legend

Comptroller	Office of the Comptroller of the Currency
FDIC	Federal Deposit Insurance Corporation
FHLBB	Federal Home Loan Bank Board
Federal Reserve	Board of Governors of the Federal Reserve System/Federal Reserve Banks
FSLIC	Federal Savings and Loan Insurance Corporation

IBF	International Banking Facility
NCUAB	National Credit Union Association Board
S&L	Savings & Loan Association
FTC	Federal Trade Commission
N/A	Not applicable

Table 3-1 (continued)
Depository Institutions and Their Regulators

	7. Access to the Discount Window	8. Deposit Insurance	9. Supervision & Examination	10. Prudential Limits, Safety & Soundness	Consumer Protection 11. Rulemaking	Consumer Protection 12. Enforcement
A. National Banks	Federal Reserve (21)	FDIC	Comptroller	Comptroller	Federal Reserve	Comptroller
B. State Member Banks	Federal Reserve (21)	FDIC	Federal Reserve & state authority	Federal Reserve & state authority	Federal Reserve & state authority	Federal Reserve & state authority
C. Insured State Nonmember Banks	Federal Reserve (21)	FDIC	FDIC & state authority	FDIC & state authority	Federal Reserve & state authority	FDIC & state authority
D. Noninsured State Banks	Federal Reserve (21)	None or state insurance fund (22)	State authority	State authority	Federal Reserve & state authority	State authority & FTC (33)
E. Savings Banks Federal Mutual	Federal Reserve (21)	FSLIC or FDIC (23)	FHLBB (27)	FHLBB	Federal Reserve & FHLBB	FHLBB
Savings Banks State Mutual	Federal Reserve (21)	FDIC or state insurance fund (23)	FDIC & state authority	FDIC & state authority	Federal Reserve, FHLBB & state authority	FDIC, state authority or FHLBB (31)
F. Savings & Loan Associations Federal	Federal Home Loan Bank & Federal Reserve (21)	FSLIC (24)	FHLBB	FHLBB	Federal Reserve & FHLBB	FHLBB
Savings & Loan Associations State	Federal Home Loan Bank & Federal Reserve (21)	FSLIC or state insurance fund (24)	FSLIC or state authority (28)	FHLBB or state authority	Federal Reserve, FHLBB & state authority	FHLBB, FTC & state authority (32)(33)
G. Credit Unions Federal	Central Liquidity Facility & Federal Reserve (21)	Credit Union Share (25)	NCUAB	NCUAB	Federal Reserve & state authority	NCUAB
Credit Unions State	Central Liquidity Facility & Federal Reserve (21)	Credit Union Share or state insurance fund (25)	State authority	State authority	Federal Reserve & state authority	State authority

Table 3-1 (continued)
Depository Institutions and Their Regulators

	Access to the Discount Window 7.	Deposit Insurance 8.	Supervision & Examination 9.	Prudential Limits, Safety & Soundness 10.	Consumer Protection 11. Rulemaking	Consumer Protection 12. Enforcement
H. Bank Holding Companies	N/A	N/A	Federal Reserve	Federal Reserve	Federal Reserve & state authority	FTC (33)
I. Foreign Branches of U.S. Banks National & State Members	N/A	N/A	Comptroller or Federal Reserve (29)	Comptroller & Federal Reserve	N/A	N/A
Foreign Branches of U.S. Banks Insured State Nonmembers	N/A	N/A	FDIC or state authority (29)	FDIC & state authority	N/A	N/A
J. Edge Act Corporation	N/A	N/A	Federal Reserve	Federal Reserve	N/A	N/A
Agreement Corporations	N/A	N/A & state authority	Federal Reserve	Federal Reserve	N/A	N/A
K. International Banking Facilities	N/A	N/A	Federal Reserve	Federal Reserve	N/A	N/A
L. U.S. Branches & Agencies of Foreign Banks—Federal	Federal Reserve (21)	FDIC (26)	Comptroller & Federal Reserve (30)	Comptroller	Federal Reserve & state authority	Comptroller
U.S. Branches & Agencies of Foreign Banks—State	Federal Reserve (21)	FDIC (26)	FDIC, state authority & Federal Reserve (30)	Federal Reserve, or state authority	Federal Reserve & state authority	Federal Reserve, FDIC, or state authority

Legend

Comptroller	Office of the Comptroller of the Currency
FDIC	Federal Deposit Insurance Corporation
FHLBB	Federal Home Loan Bank Board
Federal Reserve	Board of Governors of the Federal Reserve System/Federal Reserve Banks
FSLIC	Federal Savings and Loan Insurance Corporation

IBF	International Banking Facility
NCUAB	National Credit Union Association Board
S&L	Savings & Loan Association
FTC	Federal Trade Commission
N/A	Not applicable

Table 3-1 (continued)
Depository Institutions and Their Regulators

This Matrix provides an overview of the primary regulators of depository institutions as of January 1988. It is not intended to cover each area of regulatory responsibility in detail. Further, the Matrix and the accompanying footnotes should not be considered either a substitute for the regulations or an interpretation of them. For answers to specific questions, the regulatory agencies should be consulted.

(1) Agreement Corporations are subject to the restrictions on powers established by the Federal Reserve for Edge Act Corporations.

(2) Federal credit unions are not required to receive approval from the NCUAB before opening a branch.

(3) The establishment of federal branches and agencies is subject to the within-state branching restrictions of the McFadden Act. The establishment of state branches or agencies is regulated by state banking law. A foreign bank may not relocate any insured branch of agency within the state without the prior written consent of the FDIC

(4) While the McFadden Act prevents interstate branching by national and state member banks, banks can provide certain services on an interstate basis.

(5) While the McFadden Act's interstate branching restrictions are not applicable to insured state nonmember and noninsured state banks, state laws generally prohibit branching by out-of-state banks.

(6) As a matter of policy, the FHLBB has prohibited interstate branching by federal thrifts. Limited exceptions have been made in cases of failing institutions.

(7) Foreign banks with state or federal branches or agencies (or commercial lending companies or bank subsidiaries) are not permitted to establish a federal or state branch or agency outside their home state unless: 1) it is permitted by law in the state in which it will operate and 2) in the case of a branch, an agreement with the Federal Reserve has been entered that the nonhome-state branch will limit deposits to those permitted to Edge Act Corporations.

(8) The Comptroller must approve the merger or acquisition if the resulting bank is a national bank. However, if a noninsured bank merges into an insured bank, the FDIC must approve the merger.

(9) The Federal Reserve must approve the merger or acquisition if the resulting bank is a state member bank. However, if a noninsured bank merges into an insured bank, the FDIC must approve the merger.

(10) The FDIC must approve the merger or acquisition if the resulting bank is an insured state nonmember bank or if a noninsured bank merges into an insured bank.

(11) In addition to state authority the FDIC must approve mergers or acquisitions between insured and noninsured banks.

(12) The FDIC must approve the merger or acquisition if the resulting bank is an insured bank other than a federal savings bank.

(13) The NCUAB must approve the merger or acquisition if the resulting credit union is federally insured.

(14) The Federal Reserve supervises acquisitions made by Edge Act Corporations and Agreement Corporations. Agreement Corporations may merge as permitted by state authority.

(15) The International Banking Act of 1978 makes foreign banks that have branches of agencies in the U.S. subject to the provisions of the Bank Holding Company Act of 1956 with respect to acquisitions other than banks. Acquisitions of banks are also subject to the Bank Holding Company Act and the home-state limitations imposed by the international Banking Act.

(16) The McFadden Act prevents interstate branching by national and state member banks.

(17) The Douglas Amendment to the Bank Holding Company Act allows bank holding companies to acquire banks in other states if the state of the acquired bank specifically allows out-of-state bank holding companies to acquire in-state banks.

(18) Under the depository Institutions Deregulation and Monetary Control Act of 1980, the Federal Reserve is required to set a uniform system of reserve requirements (Regulation D) for virtually all depository institutions. Noninsured state banks which are eligible for deposit insurance may be subject to reserve requirements.

(19) Deposits at foreign branches of U.S. banks that are payable only outside the U.S. are not subject to reserve requirements.

(20) Regulation D provides that deposits at IBFs satisfying the requirements of that Regulation are exempted from reserve requirements.

Table 3-1 (continued)
Depository Institutions and Their Regulators

(21) Nearly all depository institutions in the U.S., including branches and agencies of foreign banks, have access to the discount window. These depository institutions are expected to make reasonable use of their usual sources of funds before turning to Federal Reserve Banks. For example, S&Ls and credit unions should first go to the Federal Home Loan Banks and the Central Liquidity Facility, respectively.

(22) Deposits which are not insured by the FDIC may be insured by state or state-authorized insurance funds.

(23) Deposits in federal savings banks are insured by the FDIC. Deposits in state savings banks are insured by the FSLIC. However, under the Garn-St Germain Depository Institutions Act, state savings banks which convert to federal charters may continue to have their deposits insured by the FDIC. Deposits in savings banks which are not insured by either of these federal deposit insurance agencies may be insured by state or state-authorized insurance funds.

(24) Deposits in all federal S&Ls as well as in many state S&Ls are insured by the FSLIC. Deposits in nonfederally-insured institutions may be insured by state or state-authorized insurance funds.

(25) Shares in all federal credit unions and many state credit unions are insured by the National Credit Union Share Insurance Fund, which is administered by the NCUAB Shares in some state credit unions may be insured by state or state-authorized insurance funds.

(26) Federal branches of foreign banks which accept retail deposits generally must obtain FDIC insurance. State branches of foreign banks which accept retail deposits generally must also obtain FDIC insurance if they are located in a state in which a state bank is required to have deposit insurance.

(27) The FDIC has the right to examine state savings banks which have converted to federal charter but whose deposits continue to be insured by the FDIC

(28) Federally insured S&Ls are supervised and examined by the FSLIC; nonfederally insured state S&Ls by state authority.

(29) Foreign branches of national banks are supervised and examined by the Comptroller; foreign branches of state member banks by the Federal Reserve; foreign branches of insured state nonmember banks by the FDIC; and foreign branches of noninsured state nonmember banks by state authority.

(30) Federal branches and agencies are examined by the Comptroller; state branches insured by the FDIC are examined by the FDIC and state authority; and state noninsured branches and agencies are examined by state authority. The Federal Reserve has residual examining authority over all banking activities of foreign banks.

(31) The FHLBB may share enforcement responsibility over savings banks which are members of the Federal Home Loan Bank System.

(32) The FHLBB would not have federal enforcement responsibility over state S&Ls which are neither members of the Federal Home Loan Bank System nor insured by the FSLIC

(33) Enforcement of federal consumer regulations is generally left to the FTC where the institution is not otherwise a federally supervised depository institution.

Source: Public Information Department, Federal Reserve Bank of New York.

to 1829.[2] Federal involvement in deposit insurance began more than 100 years later, in 1933, in the aftermath of the financial debacle of 1929-1933 and the large numbers of bank and thrift failures during those years.

This chapter will describe the structure and substance of regulation that has applied to thrifts. As will become clear, a central goal of depository regulation has been the maintenance of the financial health, or solvency, of depositories. This goal is usually encompassed within the domain of "safety-and-soundness" regulation. Even in the absence of explicit government deposit insurance this goal has been an important one, since the preservation of depositors' moneys in depositories has always been seen as an important social issue; with the presence of explicit government insurance of deposits, the consequences of depositories' insolvencies are yet more immediate for the regulatory process.

Because of the importance of solvency as a regulatory goal and because solvency is an accounting concept, this chapter will also present a brief primer on balance sheet accounting for depositories, particularly for thrifts. This discussion of depository accounting will yield many useful insights for understanding the regulatory process and will be a valuable frame of reference throughout this book.

First, though, a brief description of the three major types of regulation and their goals and methods is worthwhile. These distinctions are important for making sense of the complex web of regulation that envelopes depositories. They are also important for understanding the deregulation actions of the early 1980s and their consequences for the events later in the decade.

One major form of regulation is *economic regulation*. The usual justification for economic regulation is the need to curb the economic power, or market power, of the regulated entities.[3] This form of regulation usually involves some combination of limitations on prices, profits, and entry into specific lines of business. The state regulation of local electricity, natural gas, and telephone service is one familiar example of this type of regulation; federal regulation of the airline industry's prices and route structures prior to the early 1980s is another example. For depositories, economic regulation has taken many forms: usury ceilings, which limit the interest rates that depositories can charge on loans; ceilings on the interest rates that they can pay on deposits; limitations on branch locations within states and across states; and limitations on who can enter the depository business as well as limitations on the lines of business that depositories (and their subsidiaries, affiliates, and holding companies) can enter.

A second form is *safety regulation.*[4] This type of regulation is aimed at ensuring that consumers are provided with a safe, noninjurious product or service. Safety regulation usually involves limitations on the types of products and services that can be offered and the ways in which they can be produced. Federal safety regulation of autos, airlines, pharmaceuticals, and most foods are familiar examples. For depositories, this form of regulation focuses on the "safety-and-soundness" of the depository institution. Many of the economic regulatory restrictions mentioned earlier—notably, limitations on who can enter the depository business, what products and services depositories can offer, what activities their subsidiaries (and affiliates and holding companies) can undertake, and even what interest rates can be paid on deposits—have often been justified on safety-and-soundness grounds, as well as on economic power grounds. Furthermore, depositories are expected to meet minimum net worth (capital) levels and to maintain internal practices and procedures that are geared toward safe practices. For example, depositories are expected to have procedures for properly "underwriting" the loans that they make, by thoroughly checking out the capability of the borrower to repay the loan and the collateral or security (if any) offered by the borrower.

The third form of regulation involves *consumer information and protection,* either through the required provision of information or through a requirement of an absence of discrimination based on race, religion, gender, and other relevant categories. Labeling laws for foods and other products and securities disclosure and corporate reporting are familiar examples of the former type of regulation; nondiscrimination requirements in employment and sales practices are examples of the latter. For depositories, information requirements include specified information disclosures on mortgage terms and deposit terms; nondiscrimination requirements focus on lending practices and on the services (especially lending) provided to the local communities where depositories have their offices.

The Structure and Substance of Thrift Regulation

There is no easy way to describe the structure and substance of thrift regulation. No brief discussion of thrift regulation can do justice to its complexity and to the minutia that it covers. The text of the relevant federal laws and regulations cover many hundreds of pages. (Readers who chafe during the following discussion should take comfort in the fact that the regulatory structure that applies to commercial banks is even more complicated.) The major features, however, can be highlighted. Perhaps the best way is to begin with the dual federal-state chartering system.

Thrifts can hold either a state or federal charter. The charter constitutes a thrift's basic authorization to conduct business. The choice of a state or a federal charter subjects the thrift to the primary jurisdiction of a state or federal regulatory body and its attendant laws and regulations; however, this jurisdiction is not exclusive. Federally chartered thrifts are subject to state laws with respect to interstate branching, for example; and state-chartered thrifts that are federally insured (today, all are[5]) are subject to federal safety-and-soundness regulation that is linked to their deposit insurance. As was shown in Table 2-8, as of the end of 1989, 42 percent of SAIF-insured thrifts were state-chartered; they tended to be smaller than their federal brethren and accounted for only 26 percent of all assets.[6]

State regulatory powers are exercised either through a separate state agency[7] or through an agency that regulates banks as well as thrifts.[8] The state regulatory domain can include: the qualifications of who can own and operate a thrift; a thrift's loan and investment powers (i.e., the types of assets it can acquire[9]); the types and terms of liabilities a thrift can issue; the related activities it can enter[10]; and its branching locations within the state and across state lines. State regulators can appoint a conservator or receiver if a thrift fails to meet certain safety-and-soundness criteria; a thrift's insolvency is a frequent triggering criterion. The appointment of a conservator effectively removes the control of a thrift from its owners and managers, though the owners still have a claim on the assets that a conservator is supposed to "conserve." The appointment of a receiver effectively extinguishes the interests of the owners and places the thrift at the complete disposal of the regulator; it is, in essence, a withdrawal of the charter.

Until August 1989 federally chartered thrifts were subject to regulation by the Federal Home Loan Bank Board[11]; since August 9, 1989, those regulatory powers have been exercised by the Office of Thrift Supervision (OTS), a bureau of the U.S. Department of the Treasury. The domain of federal power is extensive. It includes: the qualifications for obtaining a federal charter; the types of loans and investments a thrift can make; the types of liabilities a thrift can issue; the permissible activities of the thrift and its subsidiaries, affiliates, and holding company; the limits on transactions between a thrift and its related entities (including managers, directors, and owners); the minimum required net worth level for a thrift; the general procedures that a thrift should follow for operating in a safe and sound manner; and the conditions under which the agency can appoint a conservator or receiver, effectively removing control of the thrift from its owners and managers. The major federal grounds for the appointment of a conservator or receiver are insolvency, the substantial dissipation of assets, or the operation of a thrift in an unsafe or unsound condition.[12]

The motives for the web of state and federal regulation discussed thus far are usually a combination of concerns about their exercise of economic power and concern about the preservation of the safety and soundness of thrifts.

Federal regulatory authority also extends to information disclosures related to loans and deposits and to nondiscrimination and community service requirements. These regulations apply to both state-chartered and federally chartered thrifts.

Finally, deposit insurance brings yet another layer of regulation into the process. All federally chartered thrifts are automatically covered by federal deposit insurance and hence subject to this layer of regulation. For state-chartered thrifts, a usual condition for the maintenance of their charter is their coverage under federal deposit insurance,[13] so they too are subject to this regulation. As was illustrated in Table 2-1, as of the end of 1989, all thrifts were insured by the FDIC: More than 80 percent were in the SAIF, and the remainder (the "mutual savings banks") were in the BIF. Prior to August 1989, the former group had been insured by the FSLIC, whereas the latter group had always been insured by the FDIC.

The regulation related to deposit insurance pertains partly to defining what is an insured account, who is covered, how insurance premiums (levied on the thrift) are determined and collected, and so on. But its major focus is on safety-and-soundness regulation and solvency, since the deposit insurer is exposed to potential loss from thrift insolvencies. Insurance regulation includes: the criteria for granting insurance to a thrift; the conditions and procedures for withdrawing insurance; the criteria for appointing a conservator or receiver (which has special relevance for state-chartered, but federally insured, thrifts); special limits on practices (e.g., the types of loans and investments permitted) that could endanger the safety and soundness of the thrift and thus put the insurer at risk; and the minimum net worth (capital) requirements that insured thrifts are expected to observe. The responsibility of the formal regulator (the OTS) and of the insurer (the FDIC) in the safety-and-soundness area are substantially overlapping, and in some instances the regulator has specific responsibilities (e.g., the appointment of a conservator or receiver for state-chartered thrifts; the setting of minimum net worth standards) that are, in essence, insurance regulation.[14]

The net worth requirements are at the heart of solvency as a goal of regulation. To understand the importance of net worth, and to gain greater insights into the regulatory function, it is useful to explore some simplified accounting concepts.

A Simple Accounting Framework

As was true of the description of the regulatory system, no short discussion can do full justice to the complexities of thrift accounting. Nevertheless, a simple balance sheet yields many useful insights.

Tables 3-2, 3-3, and 3-4 provide a greatly simplified balance sheet for a thrift. (The same general picture would be valid for a bank or a credit union.) The balance sheet provides a "snapshot" of the assets and liabilities of a company at a specific point in time.[15] In each table, the assets of the thrift are the loans that it has made.[16] The loans are assets from the perspective of the thrift because they represent investments; the thrift expects to receive interest and to have the principal repaid. The main liabilities of the thrift are the deposits that it takes in.[17] The deposits are liabilities or debt from the perspective of the thrift because the thrift owes this amount to the depositors; it is "liable" to them. The deposits are insured; the depositors have been given a guarantee by the deposit insurer that they will remain whole, regardless of the actions or outcome of the thrift. In practice, federal deposit insurance has always specified a maximum amount—$100,000 of principal and accrued interest, as of 1989; for the purposes of this example it will be assumed that the maximum is not relevant and that all deposits are covered.

One other aspect of these balance sheets should be noted: The example assumes that the asset and liability values listed on the balance sheets represent *market values*. This is a crucial point; discussion later in the chapter will return to it.

Table 3-2
The Balance Sheet of a Healthy Thrift, as of
December 31, 199X

Assets	Liabilities
$100 (loans)	$92 (deposits, insured)

	$8 (net worth)

Table 3-2 portrays a solvent and healthy thrift. The value of its assets exceeds the value of its deposit liabilities, so there is "something left over." This "something left over" is the net worth or capital of the thrift.[18] It is the property or financial stake of the owners of the thrift in the enterprise. For example, if the thrift were to be wound up, the assets could be liquidated (e.g., the loans could be sold to another depository institution) for $100, the depositors could be paid their $92, and there would be $8 left over for the owners. Equivalently, if the owners of the thrift wanted to sell the entire institution as a going concern to another set of owners, the latter should be willing to pay approximately $8 to the former.[19]

In Table 3-3 we see a "marginal" thrift, in which the value of the thrift's assets are worth only $92, and the thrift's net worth has vanished. The healthy thrift of Table 3-2 could become the marginal thrift of Table 3-3 if it has to "write down" the values of some of its assets because some of its borrowers default and cannot repay their loans (or they can only partially repay or require easier repayment terms)

Table 3-3
The Balance Sheet of a Marginal Thrift, as of
December 31, 199Y

Assets	Liabilities
$92 (loans)	$92 (deposits, insured)

	$0 (net worth)

Table 3-4
The Balance Sheet of a Deeply Insolvent Thrift, as of
December 31, 199Z

Assets	Liabilities
$60 (loans)	$92 (deposits, insured)

	−$32 (net worth)

or if the thrift cannot earn enough interest income from its asset portfolio to cover its overall costs of operations (of which the major portion is the cost of paying interest to its depositors) and must therefore sell some of its assets to obtain the funds to pay its costs.[20] With their net worth gone, the thrift's owners' stake in their business has also become zero.

Finally, Table 3-4 shows a deeply insolvent thrift, the assets of which have fallen greatly in value.

Insights from the Accounting Examples

There are important insights for thrift regulation that can be gained from the simple balance sheets just presented. It is important to note that *these insights apply as validly to banks and credit unions.*

1. Depositories—even those that are considered healthy—are very thinly capitalized institutions. There are two, mathematically equivalent, ways of measuring capitalization. One is to measure the ratio of the enterprise's net worth to its total assets. By this measure, the healthy thrift has an 8 percent net worth ratio. The other method is to calculate the enterprise's debt-to-equity ratio: the ratio of the enterprise's debt liabilities (in our example, the thrift's deposit liabilities) to its net worth. By this measure, the healthy thrift has an 11.5:1 debt-to-equity ratio.

Banks and thrifts are quite thinly capitalized as compared with other enterprises in the U.S. economy. This comparison can be seen in Table 3-5. As of 1985 the average for all corporations was a net-worth-to-assets ratio of 0.32 or a debt-equity ratio of 2.11:1. The net-worth-to-assets ratios for banks and thrifts were far lower; their debt-equity ratios were far higher.

The consequence of thin capitalization is that comparatively small percentage changes in the value of a thrift's assets can mean comparatively large percentage changes in its net worth. This is frequently referred to as the principle of *leverage.*

2. Under the U.S. legal system, corporations and their owners operate in a legal framework of limited liability: The financial liability of a corporation's owners is legally limited to their initial investment in the enterprise.[21] If the corporation becomes bankrupt and cannot meet its obligations to its creditors, the owners cannot be required to contribute more resources to the corporation.[22] In essence, the owners can "walk away" from the company at the time that its net worth falls to zero. Their limited liability has limited their exposure to the possible adverse or downside consequences of actions taken by the corporation.[23] In the example of the insolvent thrift of Table 3-4, the principle of limited liability means that the diminished amount of assets would impose losses (equal to the negative net worth) on the depositors, or on their insurer—but not on the owners.

At the same time, however, the owners of the corporation enjoy the full benefits of any favorable or upside consequences of the corporation's actions.[24]

One way of describing the financial position of the owners is that they have a "put option" vis-à-vis the debt liability holders: At zero or negative net worth, the owners can "put" the company to the debt holders and walk away. In the case of banks and thrifts, however, the depositors are insured (up to the insured amount). Consequently, it is primarily the deposit insurer (plus any uninsured depositors and other uninsured creditors) that is exposed to the downside consequences of the put option arrangement.

This put option framework provides useful insights for understanding the incentives for risk-taking by thrift owners (or by thrift managers on behalf of owners) under the current system of deposit insurance:

a. With the protections of limited liability, thrift owners prefer *more* risk-taking than would be true if limited liability were absent.[25] This is true because the owners will reap the upside rewards of the favorable outcomes of greater risk-taking, while limited liability limits *their* losses from the downside consequences of the unfavorable outcomes of the greater risk-taking.

b. The owners would generally prefer to operate with less net

Table 3-5
Measures of Corporate Net Worth for Major Industry
Groups, 1985

Industry	Ratio of Net Worth to Total Assets	Ratio of Debt to Equity
All industries	0.32	2.11
Agriculture, forestry, and fishing	0.32	2.12
Mining	0.45	1.21
Construction	0.28	2.52
Manufacturing	0.45	1.20
Transportation and public utilities	0.40	1.50
Wholesale and retail trade	0.29	2.49
Services	0.31	2.25
Finance, insurance, and real estate	0.26	2.90
Bank holding companies	0.08	11.07
Commercial banks[a]	0.08	11.00
Mutual savings banks	0.06	15.75
Savings and loan associations	0.03	31.12

[a] Excluding bank holding companies.

Source: U.S. Internal Revenue Service data.

worth rather than more. This is true because the owners have less to lose from the downside outcomes of risk-taking when their net worth is lower; at the limit, they have nothing to lose when their net worth is zero.[26] By keeping their investment at a minimum they thereby get the greatest potential gains from leverage. Also, the gains from risk-taking increase as net worth decreases, so the thrift owners would be inclined to take greater risks when their net worth is lower.

    c. The owners would prefer that the insurer (or the regulator acting on the insurer's behalf[27]) have *weaker* powers to curb the risk-taking by the owners. Also, the ability of the insurer to appoint a receiver (e.g., when the thrift becomes insolvent) has the effect of a call option: The insurer can "call" the enterprise from its owners. The owners' risk-taking is of greater value to them when the insurer is less able to exercise its call option.[28] Thus, it is to the owners' benefit to reduce the regulatory power of the insurer to restrict the thrift's risk-taking activity or to appoint a receiver. They might do this by providing misleading information to the insurer or by achieving, through the political process, restrictions on the insurer's regulatory powers. Also, as was true in the case of reduced net worth, since the gains from risk-taking increase as the regulatory powers of the insurer diminish, the thrift owners would be inclined to take greater risks when the insurer's regulatory powers are diminished.

Although these last two propositions have been couched solely in terms of risk-taking, the same principles apply in terms of temptations to engage in fraudulent behavior, to violate regulations, or simply to be lax or careless in underwriting loans. The lower is the thrift's net worth or the insurer's protective powers, the greater are the incentives for the thrift's owners to engage in these undesirable behaviors.

3. Because the insurer bears the downside consequences of the thrift's actions (if those consequences are substantial enough to eliminate the thrift's net worth)—regardless of whether those consequences are due to bad luck, bad judgment, carelessness, excessive optimism, excessive risk-taking, or fraud—the insurer needs to protect itself against those consequences. Among the means that the insurer could use to protect itself would be direct restrictions on the activities and powers of insured thrifts, an insistence on procedures and internal controls within thrifts that would reduce carelessness, and a requirement that thrifts maintain a minimum level of net worth.[29] The safety-and-soundness regulation discussed earlier in this chapter thus can be seen to represent the efforts of the insurer (or the regulator acting on the insurer's behalf) to protect itself. Further, these regulations are the rough equivalent of the covenants that bondholders place in their bond indentures or the restrictions that banks place in their lending agreements before they lend to a corporation, because these groups (as liabilities holders) face the same problem as the insurer: being exposed to the put option on the part of the corporation's owners.

Most of the regulatory restrictions involve "command-and-control" regulation: the regulations, in essence, instruct the thrift to avoid certain activities or to follow certain procedures. The requirements that thrifts maintain specified net worth levels, however, are different, and warrant further discussion, because they bring economic incentives into the picture.

As Table 3-2 indicates, net worth is a direct buffer or protection for the insurer. The greater is the level of net worth, the greater can be the erosion of asset value (due to any cause) before the insurer is exposed to loss. But higher net worth also has important indirect effects. As the preceding discussion indicated, lower net worth levels provide a greater incentive for thrift owners to engage in risk-taking activities. Conversely, higher net worth levels are likely to decrease risk-taking. In essence, with higher net worth levels, the thrift's owners have more of their own resources at stake—more to lose from the downside consequences of risk-taking (or of carelessness, etc.)—and so are likely to be more cautious and conservative.

There is another way of expressing this point. The net worth of a thrift is like a deductible provision in a standard liability insurance policy.[30] A deductible provision directly protects the insurer, and it is likely to have beneficial indirect effects by causing the insured party to be more cautious. Net worth has the same direct and indirect effects.

It is worth noting that another means of protecting the insurer that makes use of incentives—a premium structure that would allow premiums to vary directly with the riskiness of the thrift's activities—is not currently allowed by law. Only flat-rate premiums, which ignore risk, can be charged.[31]

4. The issue of adequate net worth and solvency should not be confused with the *liquidity* of the assets.[32] Solvency refers to the *value* of the assets relative to the liabilities; liquidity refers to the ease with which the assets can be converted into cash (so as to meet depositors' immediate withdrawal demands). The thrift of Table 3-2 might have relatively illiquid assets. But it is solvent, because the value of its assets exceeds the value of its deposit liabilities. And, so long as it can borrow against those assets (e.g., from its Federal Home Loan Bank, from the Federal Reserve, or from another lender), it would not face liquidity problems in meeting any depositors' withdrawal demands. By contrast, the thrift of Table 3-4 is insolvent because the value of its assets are far below the value of its liabilities. This would be true even if all $60 of its assets were in the form of cash sitting idly in its vaults.

Thus, liquidity should be of little concern to the deposit insurer or to the thrift or its depositors, so long as the thrift's assets provide good collateral for any borrowing that it might need to meet the liquidity demands of its depositors. Instead, it is the thrift's solvency and positive net worth that provide the assurance that the value of its assets (with the possible supplement of interim borrowing for liquidity) is adequate to meet all of the claims of its depositors.

5. Because net worth is such an important protection for the insurer and because net worth is calculated as the residual of asset values and deposit liability values, the insurer should want to receive accurate information about the *market* values of those assets and liabilities. Unfortunately, the existing accounting system—usually la-

beled generally accepted accounting principles (GAAP)—does *not* provide this information. For the most part, GAAP is based on historical book-value costs rather than on current market values. Thus, if a thrift makes a thirty-year mortgage loan for $100,000 at a fixed interest rate of 10 percent, the thrift can continue to carry that loan on its books as an asset worth $100,000 (less any repayment of principal), even if current interest rates have changed substantially. An increase in market interest rates—say, to 15 percent—would cause the *market value* of that 10 percent loan to decrease, since the higher current interest rate (or discount rate) would cause the present discounted value of the future flow of interest and principal repayment of the 10 percent mortgage to be substantially less than $100,000.[33] GAAP, however, ignores this change in value. So long as the thrift can claim that it intends to hold a loan or investment until maturity and that the likelihood of repayment of principal is not significantly in question, the thrift can carry that asset at historical book value. GAAP similarly ignores the current market values of liabilities.

It should be emphasized that GAAP's reluctance to acknowledge market value is not unique to thrift accounting. This is the general approach of GAAP for all industries, except for investment banking and pension funds (which are required to report market values).

This backward-looking accounting framework is unfortunate because it deprives the insurer of the knowledge of how much net worth protection is really available. But, to the extent that managers and thrift owners are aware of true market values (and are not themselves misled by GAAP reporting), it is the market value of net worth that will influence their incentives for risk-taking.

Further, the incentives created by this method of accounting lead to a distinct bias toward the overstatement of asset values. If the value of a loan or investment goes up in market value (as compared with its historical cost), that loan can frequently be sold, and the thrift can thereby book the gain; if the value goes down, the thrift can hide the decreased value by simply holding on to the asset and continuing to record its value at historical cost.[34]

Finally, even if the insurer has other informational means for determining the market value of a thrift's assets and liabilities, it is the formal accounting reports against which formal regulatory requirements are measured. The ability to take regulatory action when formal GAAP accounting indicates solvency for a thrift, even though other (market-based) measures indicate insolvency, becomes substantially more difficult.

In short, the absence of market value accounting places the insurer at a substantial disadvantage. At best, the insurer should be compensating for its informational disadvantage by maintaining tighter regulatory restrictions elsewhere and by requiring higher GAAP net worth levels. At worst, the insurer is exposed to greater losses.

6. Though market value accounting would be a great improvement over the current GAAP framework, even that is not adequate information for the protection of the insurer because it is too static. The insurer also needs "dynamic" accounting information—in essence, "what if?" simulations—to gain information about its exposure to the risks that are *already* embedded in the assets and liabilities of the thrift.[35] This would include information about the possible value consequences (in market value terms) of changes in interest rates or in general economic activity (which might affect loan default rates).

This need can be demonstrated by the following example: Suppose, first, that the solvent thrift of Table 3-2 had all of its $100 of assets invested in short-term Treasury bills, while its $92 of deposits also were short term. Second, suppose instead that the solvent thrift had its $100 of assets entirely invested in thirty-year fixed interest rate Treasury bonds at current interest rates, while its $92 of deposits were still short-term. For "static" accounting purposes, the two thrifts would each be accurately portrayed by the balance sheet of Table 3-2, even if market values were used; static accounting could not distinguish between them. But the first thrift is clearly exposed to much less risk from interest rate fluctuations than is the second. If interest rates go up, the first thrift will have to pay higher interest on its deposits, but it will earn higher interest rates on the new Treasury bills that it will buy as its current T-bills mature; it will suffer no serious erosion of earnings, and its asset and liability market values will remain relatively unchanged. By contrast, the second thrift is exposed to much greater risk from an increase in interest rates since it too would have to pay higher rates to its depositors, but its income from its long-term Treasury bonds would remain unchanged. It will suffer a decrease in net earnings or losses, and its asset market value will diminish relative to its liability market value, eroding its net worth.

The insurer needs to know about these risks and needs to be able to distinguish between the first thrift and the second thrift. Static accounting does not provide this information. Only a dynamic framework, with enough information to allow the "what if?" simulations, can satisfy the insurer's needs.

7. The focus on accuracy in accounting information is important for another, related reason. Thrift owners have a strong incentive to try deliberately to *overstate* the value of the thrift's assets. If the thrift would otherwise be shown to be insolvent (and the appointment of a receiver would be imminent), overstating the value of the assets would achieve apparent solvency. If the thrift would otherwise be solvent but failing to achieve a regulatory minimum required net worth level (which could thereby bring regulatory restrictions or just unfavorable publicity), overstating would achieve apparent regulatory compliance. Further, even if no regulatory minimum were currently being breached, the overstatement of asset values would allow greater expansion (greater leverage) than would otherwise be the case; or it would allow owners to issue dividends to themselves (i.e., to pay assets to themselves) that apparently would not cause the thrift's net worth to fall below required levels.

All of these actions, however, would, place the insurer at greater risk because the thrift's net worth in actuality would be less than the overstated assets would indicate. Thus, the thrift owners and the insurer are likely to be constantly at odds over the statement of asset values.[36]

It should be noted that paying excessive dividends is not the only way that owners can "milk" a thrift and drain assets from it, to the detriment of the insurer. Loans to owners (or their companies, affiliates, customers, suppliers, relatives, friends, etc.) at favorable rates, for speculative ventures, or under other circumstances where the loans are unlikely to be repaid, is another way. Purchasing goods or services from owners (or their companies, etc.) at excessively *high* prices is yet another way. Accordingly, the insurer needs to restrict, and closely scrutinize, any transactions between the thrift and its owners. And, indeed, such restrictions are part of the regulatory landscape.

8. Once a thrift has reached a state of insolvency represented by Table 3-4, the insurer's losses are roughly equal to the market value negative net worth of the thrift. If the thrift is liquidated, the insurer must pay the depositors their $92 and sell the assets for their market value of $60, for a net loss of $32.[37] If, instead, an acquirer can be found for the thrift—say, because it has "franchise value" or "going concern value" that can be preserved if the thrift is transferred as a going concern to new owners but is lost if the thrift is liquidated—the new owners will not be willing to take over the operation of the thrift unless additional assets are provided so as to bring the thrift's total

assets into rough equality with its liabilities.[38] Again, the insurer faces an approximate cost of \$32. Only a reneging on the guarantee to depositors would allow the insurer to avoid incurring these costs.

It clearly behooves the insurer to protect itself and prevent thrifts from reaching insolvency. Unfortunately, as the following chapters demonstrate, the FSLIC did not adequately protect itself in the 1980s, and it allowed hundreds of thrifts to reach lesser and greater approximations to Table 3-4.

Notes

1. One such effort, for commercial banks only, can be found in Bloch (1985, p. 164); see also Figure 9-1.
2. See Calomiris (1989a; 1989b).
3. In practice, though, much economic regulation has had the effect, and often the intention, of protecting otherwise competitive industries from the effects of competition; see, generally, Stigler (1971), Posner (1974), Peltzman (1976), Weiss and Klass (1981), and Weiss and Klass (1986).
4. This is a narrower form of a broader category of regulation that is frequently described as "social regulation" or "health-safety-environment regulation"; see White (1981).
5. Prior to the late 1980s some state-chartered thrifts were covered by state-sponsored insurance funds. A number of these insurance funds failed—the failures of the Ohio and Maryland funds in 1985 probably gained the most notoriety—and all thrifts today are federally insured; see Kane (1989).
6. Of the BIF-insured thrifts (the "mutual savings banks"), more than 90 percent by number are state-chartered, and they account for more than 80 percent of the MSBs' assets.
7. This is true, for example, for California and Texas.
8. This is true, for example, for New York.
9. The FIRREA greatly restricted state powers in this respect, causing much greater conformity with the limitations on federally chartered thrifts.
10. See Note 9.
11. The authorizing statute was the Home Owners' Loan Act of 1933 and its subsequent amendments.
12. Other grounds include the failure of a thrift to meet liquidity demands made upon it or its violation of a cease-and-desist order. The FIRREA provided additional grounds; see Chapter 11.

13. See Note 5.
14. Prior to August 1989, the Federal Home Loan Bank Board was both the regulator and insurer of the FSLIC-insured thrifts, so the distinction between regulator and insurer was much less important.
15. The balance sheet is linked, through double-entry bookkeeping, with an income statement (also known as a profit-and-loss statement). The latter traces the flows of revenues and costs during a specified period and indicates whether the enterprise earned a profit or loss during that period. The income statement provides an important linkage between the balance sheet snapshot at the date immediately preceding the period covered by the income statement and the balance sheet snapshot at the end of this period.
16. In practice, other assets would include cash, securities, ownership interests in subsidiaries, ownership interests in other enterprises, and direct ownership of land and buildings (including the thrift's premises for operations).
17. In practice, other liabilities would include bonds issued, other forms of debt issued, and moneys owed to trade creditors.
18. Newspaper accounts of the thrift crisis have frequently referred to a thrift's capital as "cash" or "cash capital." This is incorrect and misleading. Cash is an asset. Capital (as its synonym, *net* worth, implies) is simply the residual of the value of the enterprise's assets, less the value of its liabilities. It is the owners' *stake* in the enterprise.
19. If the thrift has "franchise value" or "going concern value" that is not adequately represented by the market values of the listed assets and liabilities (e.g., the thrift has a good reputation and can be expected to earn higher profits in the future that cannot be attributed to the existing assets and liabilities) the new set of owners should be willing to pay somewhat more than $8 for the thrift—say, $10. This $10 represents the new owners' net worth in the thrift. To keep the assets, debt liabilities, and net worth in an arithmetic linkage with each other, the extra amount paid (the $2) would be listed as an asset—"goodwill"—on the thrift's balance sheet under its new owners.
20. Most depositors do not withdraw the interest that they earn but instead let it accumulate in the thrift. This "interest credited" adds to the thrift's liabilities, which could cause net worth to erode even if the thrift did not sell assets. Or, if the thrift attracted new deposits or borrowed in some other form in order to obtain the cash to pay expenses (rather than either keeping the cash as an asset or using it to invest in new assets), this too would erode net worth.
21. This is true regardless of whether the owners are the original investors in the corporation or are subsequent shareholders who bought their interests from an earlier set of owners.

22. This assumes an absence of negligence or fraud on the part of owners who are managers or directors and who might thereby be legally liable for civil or criminal penalties.

23. Mutually organized thrifts are nominally owned by their depositors. In practice, the depositors have only a tenuous claim on the net worth of a thrift—even in the event that a thrift converts from mutual to stock ownership—and would certainly bear no financial responsibility for any negative net worth that might arise.

24. The managers of a mutual thrift have some of the same incentives as owners. Though they do not enjoy the full benefits of upside outcomes, their jobs, salaries, and perquisites are certainly linked to favorable outcomes, while they bear no financial responsibility for the negative net worth that might arise from downside outcomes. Their losses are limited to the loss of their jobs. The same set of incentives apply to professional managers of publicly traded stock thrifts, except that any stock or stock options owned by them would add to their upside gains.

25. This is frequently described as the problem of "moral hazard" behavior.

26. If net worth is negative, however, the owners do not get the full benefits of the upside of risk-taking, since some of this benefit will be absorbed by the "filling in" of the thrift's negative net worth.

27. In the following discussion, all references to "the insurer" also encompass the regulator acting on behalf of the insurer.

28. See Brumbaugh and Hemel (1984).

29. A more complete discussion of the potential protections available to the insurer appears in Chapter 10.

30. It is not identical, in the sense that an explicit deductible in an insurance policy is a predetermined amount, whereas the entire net worth of a thrift—even if it is above the required minimum—serves as a buffer for the deposit insurer.

31. As of 1989, these premiums were 8.33 cents per $100 of deposits for banks and thrifts in the BIF and 20.83 cents per $100 of deposits for thrifts in the SAIF. The FIRREA established a schedule of rates for future years, which is shown in Table 9-1.

32. See Note 18.

33. If interest rates jumped to 15 percent the day after that 10 percent mortgage were made, the present discounted value of that mortgage would be only $70,000 (if it were paid off over the full thirty years). Because mortgages typically are repaid in less than thirty years (because homeowners move or because interest rates might fall far enough to make refinancing worthwhile), the decline in value for the expected life of the mortgage would be somewhat less—but still substantial.

34. This practice of selling the "winners" and hiding the "losers" is usually described as *gains trading*.

35. This need for dynamic information would also apply to any "off-balance-sheet" items.
36. In principle, the understatement of liability values would have the same result on apparent net worth. In practice, the statement of liabilities seems less vulnerable to manipulation.
37. Technically, a liquidating receiver would be appointed, who would sell the assets and return the funds to the insurer.
38. If the thrift's franchise value is not listed as an explicit asset before the transaction, then the extra assets provided by the insurer need not completely fill the negative net worth "hole," since the franchise value will then become an explicit goodwill asset.

PART TWO

Prelude, Deregulation, and the Debacle

CHAPTER FOUR

Developments before 1980

Prior to 1980 the thrift industry was largely a sleepy industry that made home mortgages and took in passbook savings deposits. It was an important part of the larger American policy effort to encourage home ownership. It had suffered financial reverses during the financial debacle of the early 1930s; but the federal chartering, regulation, and deposit insurance that were established in the 1930s provided a solid base for industry expansion and prosperity. The industry was safe and stable.

There was, however, one flaw in this pattern: Thrifts were taking in short-term deposits but making long-term, fixed-interest-rate mortgage loans. If interest rates increased significantly, they would be squeezed. This problem first arose in the mid-1960s. The Congress applied a regulatory solution that seemed to work, and the industry (and the Congress) hoped that the problem would go away. Unfortunately, it did not, and the mounting inflation rates of the late 1970s re-created the problem with a vengeance.

Establishing the Thrift Regulatory Structure

The stock market crash of 1929 and the financial and economic decline of the following four years was a watershed for federal regulation of financial markets and institutions. The failures of thousands of banks and thrifts during the 1929-1933 period, and the consequent large losses by depositors, created a political demand for expanded federal regulation and for federal deposit insurance. The Banking Act of 1933 gave expanded regulatory powers over commercial banks to the Office of the Comptroller of the Currency and to the

Federal Reserve, and it established the Federal Deposit Insurance Corporation to provide insurance to depositors in these banks. The maximum insured amount per deposit in 1933 was $2,500.

Simultaneously, other legislation created, for the first time, a significant federal regulatory structure for thrifts. The Federal Home Loan Bank Act of 1932 created the Federal Home Loan Bank Board and a system of twelve regional Federal Home Loan Banks that would serve as wholesale lenders to the thrift industry.[1] The FHLBs could borrow in national credit markets at favorable rates (because they were federally sponsored entities) and relend these moneys to their thrift industry members.

The FHLBs, though, were (and, as of 1990, continue to be) hybrid institutions. Each of the twelve FHLBs was "owned" by its thrift members: They were required to invest capital in their local FHLB; they owned the stock of that FHLB; they elected approximately two-thirds of the directors of each FHLB; and they received dividends from the profits earned by their FHLB on the latter's lending operations. At the same time, however, the FHLBs were subject to regulation, oversight, and supervision by the Bank Board in Washington, and the Bank Board appointed approximately one-third of the directors of each FHLB.

In the next year, the Home Owners' Loan Act (HOLA) of 1933 established for the first time a federal charter for thrifts and gave the chartering and regulatory authority for these federal charters to the Bank Board.[2] The chartering structure for thrifts was thereby made parallel to the dual structure that applied to banks: Thrifts could choose to have either a state charter or a federal charter.

Finally, in 1934 the National Housing Act created the Federal Savings and Loan Insurance Corporation to provide a deposit insurance system for thrifts that was comparable to the FDIC's system for banks. The FSLIC, however, was not established as a separate agency, but, was lodged instead within and under the authority of the Bank Board. All federally chartered thrifts were automatically covered by FSLIC insurance; state-chartered thrifts had the option to seek coverage by the FSLIC. The maximum insured deposit was $5,000 in 1934; the maximum deposit for FDIC-insured banks was raised to the same amount in that year. In future years, all increases in the maximum insured amount occurred simultaneously for both insurance funds. Table 4-1 shows the time sequence of increases in the insured amount. The table also shows that the insured amount has increased in real (inflation-adjusted) as well as nominal terms.

One feature concerning the establishment of the insurance funds is worth noting: The Congress, from the beginning, ignored insurance premium levels that historical experience had indicated would be necessary for the long-run solvency of the funds; instead, the Congress chose lower levels of premiums. This decision was partly based on the belief that tighter federal regulatory procedures would lessen future losses; and it was partly based on the political unwillingness of banks and thrifts to pay the higher premiums.[3] The Congress apparently did not even consider the possibility that premium levels might vary for individual banks or thrifts, depending on their risk profiles. The *unwillingness* of the Congress in the 1930s to treat deposit insurance as a phenomenon that required the application of *insurance* principles was an unfortunate precedent that would continue to be followed through the 1980s.

The structural consequence of these Acts was a variegated pattern of thrift chartering, regulation, and insurance. Federally chartered

Table 4-1
FDIC and FSLIC Deposit Insurance: Maximum
Insured Deposit Amounts, 1934–1989

Years[a]	Maximum Insured Deposit Amount	Maximum Insured Deposit Amount Adjusted for Inflation[b]
1934–1949	$5,000[c]	$30,800 – $17,300
1950–1965	10,000	34,200 – 26,150
1966–1968	15,000	38,150 – 35,500
1969–1973	20,000	45,800 – 37,100
1974–1979	40,000[d]	66,850 – 45,400
1980–1989	100,000	100,000 – 66,450

1980 dollars

[a] As of December 31 of the years specified.

[b] In 1980 dollars, using the consumer price index to adjust for inflation; the two numbers in the column represent the values for the beginning and ending years to which the nominal dollar amount applied.

[c] Insured amount was $2,500 for the first half of 1934.

[d] Coverage of $100,000 for time and savings deposits of local and state governmental units was provided in 1974 and for IRA and Keogh accounts of individuals in 1978.

Sources: FDIC and CEA data.

thrifts were required to become members of their local FHLB and to be insured by the FSLIC. State-chartered thrifts could choose to remain aloof from all federal involvement, or they could join their local FHLB only, or they could apply to obtain FSLIC insurance coverage.

As a further complication in this pattern, one group of thrifts— the state-chartered mutual savings banks that were found largely in the Northeast (but also in the state of Washington)—was insured by the FDIC. The MSBs were generally considered to be different from the majority of thrifts. The majority had their origins squarely in residential mortgage finance. They were usually known as a savings and loan association, or S&L, though in some areas they were referred to as a building and loan association—an even more direct reference to their home mortgage roots. The MSBs, though also involved in mortgage finance, usually had authority for a wider range of consumer loans. Their origins lay more as places that were designed to encourage the urban poor and recent immigrants to save their funds. It is these different origins and focuses that led to the inclusion of MSBs with commercial banks within the FDIC insurance system,[4] while S&Ls were channeled into the FSLIC. As was noted in Chapter 2, the MSBs largely escaped the debacle of the 1980s; consequently, the discussion of the thrift industry in this and succeeding chapters will largely ignore them.

Though a patchwork pattern of thrift chartering, regulation, and insurance emerged from the legislation of the 1930s, the policy intent underlying these actions was clear. Thrifts were to be a major vehicle for providing home mortgage finance. The lending powers of the thrifts were restricted largely to making home mortgage loans. The FSLIC would provide assurance for their depositors and thereby provide thrifts with a stable, low-cost source for most of their funds. The FHLBs were to provide thrifts with an additional source of low-cost financing and also serve as a lender-of-last-resort for thrifts' short-run liquidity needs. The FSLIC and the FHLBs served purposes that were similar to those provided by the FDIC and the Federal Reserve for commercial banks. The thrift structure was clearly intended to be similar to the structure for banks—except that thrifts were expected to be focused on housing.

This structural encouragement for a group of specialized housing lenders—which were largely restricted to housing loans but, in return, were protected and coddled—reflected an important strain of political thinking in the 1930s, which has persisted to the present day. The promotion and encouragement of home ownership was an important social goal, and many (if not most) policymakers believed that a specialized set of institutions with an earmarked financing stream was necessary for the provision of low-cost housing finance. Other candidates for housing finance were life insurance companies and commercial banks. But life insurance companies were unlikely to be involved in the origination of mortgages; and commercial banks were frequently unaccustomed to long-term lending for housing. Instead, banks were accustomed to short-term commercial lending to business customers. This was consistent with one strand of macroeconomic (and political) thought at the time—the "real bills" doctrine—which argued that macroeconomic stability could be ensured by a banking policy that permitted banks to lend to businesses only for short-term purposes to accommodate the latter's needs to finance orders and inventories. During the 1920s the Comptroller of the Currency had restricted the percentages of portfolios that could be devoted to long-term mortgages by national banks. Further, many other countries had developed or were developing specialized residential lending institutions (e.g., Britain's building societies).

Also, the promotion of residential finance through the promotion and strengthening of the thrift system was just one of a number of encouragements for home ownership that the American political system would create at the federal level in the 1930s and in later decades.[5] The Federal Housing Administration (FHA), created in 1934 by the HOLA, provided mortgage insurance (guaranteeing mortgage payments to the lender) for home buyers who did not have the funds for a sufficiently large downpayment (e.g., 20 percent of the sales price) to meet the lender's standards. The Servicemen's Readjustment Act of 1944, frequently called the "GI Bill of Rights," authorized the Veterans Administration (VA) to offer a similar program for veterans. The Federal National Mortgage Association (Fannie Mae), created in 1938, was designed to borrow funds and use them to buy mortgages from lenders and originators, thereby increasing the supply of funds for housing finance. The Federal Home Loan Mortgage Corporation (Freddie Mac), created in 1970, had the same function. The Government National Mortgage Association (Ginnie Mae), established in 1968, provided an extra layer of government guarantees for FHA and VA mortgages, making it easier for lenders and originators to sell these mortgages to other investors (e.g., insurance companies and pension funds) in a secondary market. In the 1970s, and especially in the 1980s, the efforts of Fannie Mae, Freddie Mac, and Ginnie Mae to develop "pass-through" securities that represented claims on bundles of home mortgages—and then to develop more complex restructurings of these securities—greatly expanded the secondary market for mortgages and thereby increased even further the supply of funds going into mortgages. And the federal income tax code, allowing home owners to deduct mortgage interest payments and local property taxes from their incomes when computing their income taxes, provided a sizable subsidy for home ownership.

Postwar Growth and Prosperity

In the policy environment just described, the thrift industry grew and flourished, especially in the post-World War II climate of an expanding

Table 4-2
Numbers of Thrifts and Their Assets, 1930–1979

	All Thrifts[a]		FSLIC-Insured Thrifts	
	Number	Assets (billions)	Number	Assets (billions)
1930	12,369	$19.3	—	—
1935	10,726	16.9	1,117	$0.7
1940	8,061	17.6	2,277	2.9
1945	6,681	25.7	2,475	6.1
1950	6,521	39.3	2,860	13.7
1955	6,598	69.0	3,544	34.2
1960	6,835	112.1	4,098	67.4
1965	6,691	187.8	4,508	124.6
1970	6,163	255.2	4,365	170.6
1975	5,407	459.3	4,078	330.3
1979	5,147	742.4	4,039	568.1

[a] Includes mutual savings banks.

Source: FHLBB data.

U.S. economy, rising incomes, and a mobile population. Table 4-2 chronicles this growth for thrifts in general and for FSLIC-insured thrifts in particular. Between 1945 and 1965, thrift assets approximately doubled every five years. The pace of growth slackened considerably between 1965 and 1970 but then quickened in the 1970s.

This postwar period was clearly the heyday of the thrift industry. Business was good; profits were healthy. One observer of the thrift industry described it as the world of "3-6-3": Thrifts could take in money at 3 percent interest on deposits; they could lend it out at 6 percent interest on mortgage loans; and thrift executives could be on the golf course by 3:00 in the afternoon.[6]

Thrifts also benefitted from the legal and policy restrictions by federal and state regulatory agencies that hindered competition among thrifts. These restrictions (which usually applied at least as stringently to commercial banks[7]) included: (1) a reluctance at both the federal and state levels to grant new charters to de novo entrants or even to grant applications for new branches to incumbent thrifts, where the new branch would encroach on another incumbent's territory; (2) limits by some states on the within-state geographic regions in which branching was permissible under any circumstances; (3) restrictions by virtually all states that made interstate branching impossible; and (4) a federal regulation that limited a thrift to making mortgage loans on properties that were no greater than fifty miles from the thrift's home or branch offices. (In 1971 this last limit was expanded to 100 miles; it was later abolished.) A justification that was frequently offered for these restrictions was that excessive competition among banks had been a contributory cause to the widespread failures of the early 1930s; hence, banks and thrifts had to be protected and prevented from competing too vigorously with one another.[8]

Further, the thrifts' primary asset—the residential mortgage—was a relatively "safe" product. Defaults among residential mortgage borrowers were rare, especially in the expanding economy and rising real estate markets of the postwar era. Consequently, insolvencies among thrifts that activated the insurance guarantee and the expenditure of FSLIC funds were comparatively rare. Table 4-3 shows the average numbers of thrift and bank failures—instances in which an insolvent bank or thrift was disposed of, either through liquidation or through placement with an acquirer—per year through 1979.

Table 4-3
Disposals[a] of Insolvent Thrifts and Banks, 1934–1979

	FSLIC[b]			FDIC		
	Number of Disposals	Disposals as a Percentage of All Insured Thrifts	Disposed Assets as a Percentage of All Thrift Assets	Number of Disposals	Disposals as a Percentage of All Insured Banks	Disposed Assets as a Percentage of All Bank Assets
			Annual Averages			
1934–1939	2.2	0.10%	0.13%	52.5	0.38	0.10
1940–1949	2.7	0.12	0.14	10.5	0.08	0.03
1950–1959	0.4	0.01	0.01	2.7	0.02	0.01
1960–1969	4.3	0.10	0.14	4.4	0.05	0.08
1970–1979	4.3	0.10	0.11	7.6	0.05	0.08

[a] Instances in which an insolvent thrift or bank was disposed of, either through liquidation or through placement with an acquirer.
[b] Includes supervisory cases

Source: Barth, Feid, Reidel, and Tunis (1989).

FSLIC-insured thrifts were not appreciably more troublesome for regulators and the insurance fund than were FDIC-insured banks. In only one year (1941) did the number of thrift failures reach as high as thirteen.

The prevailing culture and ethos of the thrift industry during this period was that of the small, mutual thrift. In 1955 only 9.6 percent of FSLIC-insured thrifts (accounting for 11.1 percent of FSLIC-insured thrifts' assets) were stock companies. As late as 1976, only 15.1 percent of FSLIC-insured thrifts (accounting for 22.1 percent assets) were stock companies. The average size thrift in 1976 had only $81 million in assets; almost half of all thrifts (46.3 percent) had less than $25 million in assets. The attitude that many thrift executives had about their business could almost be described as a "calling": After all, they were actively involved in promoting home ownership and encouraging thrift. It was no accident that Jimmy Stewart's George Bailey ran a small savings and loan association in Frank Capra's 1946 film *It's a Wonderful Life.*

The Flaw: Borrowing Short and Lending Long

There was, however, one potential flaw in this scheme. The bulk of the assets of the thrifts were long-term, fixed-interest-rate mortgage loans—typically, thirty-year loans. Thrifts were funding these loans through short-term liabilities—typically, passbook deposits that could be withdrawn at short notice. The usual phrase for this pattern is that they were "borrowing short and lending long."

A thrift with these assets and liabilities would find that its interest *income* from its investments (the fixed-rate mortgages) would respond only slightly to changes in the general level of interest rates. Simultaneously, however, it would find that its interest *costs* (on its passbook and other short-term deposits) were quite sensitive to the general level of interest rates. So long as interest rates stayed stable or fell, or rose only gradually, thrifts could—and did—make healthy profits.

If interest rates rose sharply, however, a thrift with this balance sheet would be in serious difficulty. Its income would be tied to its portfolio of fixed-rate, long-term mortgages, made in earlier years at lower interest rates.[9] If it tried to keep its costs down by not raising the interest rates it paid on its deposits (to current market rates), the depositors would withdraw their deposits in favor of other investments that were paying market rates, and the thrift would have to sell it mortgages at a loss. If, instead, the thrift were to raise the interest rates it paid so as to retain those deposits, it would run operating losses, since its costs would exceed its income. Either way, losses were unavoidable.

This potential flaw first became a serious reality in the mid-1960s. As the United States stepped up its involvement in the Vietnam War in 1965 and 1966, interest rates began to rise. Table 4-4 shows the monthly progression of short-term interest rates between 1964 and 1966. As can be seen, interest rates rose moderately toward the end of 1964, rose again toward the end of 1965 and early 1966, and continued to rise in the summer and early fall of 1966. Thrifts began to feel squeezed.

Congress responded to their plight by passing the Interest Rate Control Act in September of 1966. This Act for the first time placed ceilings on the interest rates that thrifts could pay on deposits. These ceilings had previously been applicable (beginning in 1933) only to commercial banks and had been administered by the Federal Reserve (under "Regulation Q"). The 1966 Act gave the Bank Board the power

Table 4-4
Interest Rates on Three-Month Treasury Bills,
1964–1966

	1964	1965	1966
January	3.53%	3.83%	4.60%
February	3.53	3.93	4.67
March	3.55	3.94	4.63
April	3.48	3.93	4.61
May	3.48	3.90	4.64
June	3.48	3.81	4.54
July	3.48	3.83	4.86
August	3.51	3.84	4.93
September	3.53	3.91	5.36
October	3.58	4.03	5.39
November	3.62	4.08	5.34
December	3.86	4.36	5.01

Source: CEA data.

to set the ceilings for thrifts, in coordination with the Federal Reserve. Thrifts, though, were allowed (by statute) to pay a slightly higher interest rate than could banks; a justification offered at the time was that this differential would offset the disadvantage that thrifts experienced because they could not offer their deposit customers as complete a range of other financial services (e.g., checking accounts and consumer loans, such as car loans) as could banks. This differential was initially set at 3/4 percent (75 basis points), then narrowed to 1/2 percent (50 basis points) in 1970, and then narrowed again to 1/4 percent (25 basis points) in 1973. Table 4-5 shows the time pattern of interest rate ceilings that were applied to passbook deposits for banks and thrifts. Other (higher) ceilings applied

Table 4-5
Interest Rate Ceilings on Passbook Deposits and
Treasury Bills Rates, 1964–1966

Period	Bank Passbook Ceiling	Thrift Passbook Ceiling	Three-Month Treasury Bill Interest Rate[a]
Sept.–Dec. 1966	4%	4.75%	5.28%
1967	4	4.75	4.32
1968	4	4.75	5.34
1969	4	4.75	6.68
1970	4.50	5	6.46
1971	4.50	5	4.35
1972	4.50	5	4.07
Jan.–June 1973	4.50	5	6.13
July–Dec. 1973	5	5.25	7.93
1974	5	5.25	7.89
1975	5	5.25	5.84
1976	5	5.25	4.99
1977	5	5.25	5.27
1978	5	5.25	7.22
Jan.–June 1979	5	5.25	9.36
July–Dec. 1979	5.25	5.20	10.72

[a] Average for the period.

Sources: Federal Reserve and CEA data.

to deposits with longer maturities (e.g., three months, six months, one year, etc.). For comparison, Table 4-5 also shows the average interest rates on Treasury bills for the same periods.

The interest ceilings did not halt all competition among thrifts and banks for deposits, but it did take the "price" (i.e., interest rates) out of the picture. Consequently, thrifts and banks turned to nonprice forms of competition. Prizes, such as toasters and tennis rackets, for opening an account became common. Federal regulators found that they then had to set limits on the monetary value of the prizes. Convenience of service—provided by longer hours or a more extensive branch network[10]—or not charging for some incidental services were other forms of non-price competition. These forms of nonprice competition, though, were not as costly to banks and thrifts (nor as valuable to their depositors) as interest rate competition would have been.

The Congress' Regulation Q "patch" therefore did solve the thrifts' interest rate squeeze problems for the remainder of the 1960s and most of the 1970s.[11] Most thrift depositors had few alternative investments that offered a comparable level of safety, liquidity, and convenience available to them, except for deposits in other thrifts and banks (which, of course, were also subject to the ceilings).[12] In 1970, as a way deliberately to make switching to Treasury bills more difficult for thrift depositors, the Treasury increased the minimum denomination of Treasury bill, which had previously been $1,000, to $10,000;[13] at the time, the average deposit amount in a thrift was only $3,045.

Depositors in commercial banks, however, tended to be more financially sophisticated, to have larger deposits, and thus to have more alternatives. These depositors included other U.S. financial institutions, U.S. corporations, and overseas banks and companies. Their alternatives for investing their money—notably Treasury bills, the short-term obligations of other governmental entities, and the short-term obligations of high-quality corporations ("commercial paper")—would have meant deposit withdrawals and "disintermediation." Consequently, for most of the 1970s the Federal Reserve and the Bank Board engaged in a delicate effort of setting interest rate ceilings on various sizes and maturities of deposit, in order to balance the commercial banks' needs to prevent the departure of their depositors and the thrifts' needs to avoid an earnings squeeze. At the times when the general level of interest rates were diverging more widely from the basic Regulation Q levels (see Table 4-5), this task was appreciably harder.

One route that could have eased the thrifts' dilemma over the longer term was refused to them. A few states (notably California and Wisconsin) had authorized their state-chartered thrifts to offer mortgages with interest rates that were not fixed but that varied with an index keyed to other interest rates. These mortgages, which were called "variable rate mortgages" at the time (but are now usually referred to as adjustable rate mortgages, or ARMs), would have allowed a thrift's income to rise at times of rising interest rates and thus would have moderated earnings squeezes at those times. A version of the ARM had been the standard mortgage instrument for British building societies since the nineteenth century. Federally chartered thrifts in the United States, however, were not allowed to offer ARMs.

A number of times during the 1970s the Bank Board began formulating regulations that would have permitted federally chartered thrifts to offer ARMs. Each time, Congressional leaders indicated that they strongly opposed the initiative, largely on consumer protection grounds. They apparently feared that thrifts would act arbitrarily in raising the interest rates that locked-in mortgage borrowers would have to pay. Each time the Bank Board backed down.[14]

Accordingly, as the thrift industry entered into the late 1970s the thirty-year fixed-rate mortgage was still the standard asset for the industry, and the industry was still borrowing short and lending long. The recommendations of a number of study groups and commissions[15] during the 1970s—that Regulation Q be phased out, that thrifts be allowed to offer ARMs, and that they be allowed to diversify into other activities—were ignored. The industry, its regulators, and the Congress clung to Regulation Q as the solution to the industry's problem.

Regulation Q, however, only delayed the day of reckoning for the thrift industry. In the late 1970s, as interest rates again rose—this time much more sharply—the thrifts again faced a set of choices that meant unavoidable losses, but Regulation Q could no longer save them.

Notes

1. For a more thorough discussion of the development of the Federal Home Loan Bank Board and the Federal Home Loan Bank System, see Kendall (1962), Bloch (1963), Friend (1969), Marvell (1969), Woerheide (1984), and Adams and Peck (1989).
2. By contrast, federal charters for commercial banks were first made available seventy years earlier, by the National Currency Act of 1863 and

the National Bank Act of 1864. Only in 1934, however, did the Federal Credit Union Act make federal charters available for credit unions.

3. See Barth, Bradley, and Feid (1989) and Barth, Feid, Reidel, and Tunis (1989).

4. See National Association of Mutual Savings Banks (1962).

5. See, generally, Tuccillo (1983) and Bosworth, Carron, and Rhyme (1987, Ch. 3).

6. This description is attributed to Maurice Mann by Kane (1983, p. 56).

7. See Peltzman (1965).

8. This notion has since been discredited; see Peltzman (1965) and Flannery (1985).

9. Selling these mortgages and reinvesting the proceeds in new, higher-interest mortgages would not be a solution to this dilemma. After interest rates have risen, the low-interest mortgages could only be sold at a loss, reflecting the discounted (at a higher discount rate) stream of the future earnings from these low-interest loans.

10. See White (1976).

11. As Kane (1985, Ch. 4) points out, however, the secular rise in interest rates during this period meant that the market value of their mortgage assets were falling, reducing (or possibly, as Kane argues, eliminating) their net worth.

12. Some state-chartered thrifts, though—notably in Maryland and Ohio—were insured by state-sponsored insurance funds and were not covered by the Regulation Q ceilings.

13. See Kane (1970).

14. See Strunk and Case (1988, Ch. 5).

15. See Jacobs and Phillips (1972), Phillips and Jacobs (1983), and White (1986).

CHAPTER FIVE

The Interest Rate Squeeze and the Policy Responses, 1979–1982

With thrifts committed to a pattern of borrowing short and lending long, they could not afford to pay substantially higher interest costs on their deposits for any extended period of time. Regulation Q had largely eased the thrifts' dilemma in the late 1960s and early 1970s, but the growing sophistication of financial markets and the sharply rising interest rates of the late 1970s and early 1980s effectively terminated Regulation Q's salving powers.

In the early 1980s the federal government and a number of states provided a new policy response: economic deregulation for thrifts. Unfortunately, the need simultaneously to strengthen the safety-and-soundness regulatory system was not recognized until significantly later—when far too much financial damage had already been done.

The Interest Rate Squeeze

The sharp rise in interest rates between 1978 and 1981 is documented in Table 5-1. Short-term interest rates, which had been below 6.5 percent through May of 1978, were about 9 percent by the end of that year. They hit double digits in September of 1979 and mostly stayed in double digits for the next three years.

The Regulation Q ceiling through all of 1978 and the first half of 1979 was 5.25 percent on S&L passbook accounts; in July 1979 it was raised to only 5.50 percent. These ceiling rates were far below the market rates for short-term investments in 1978 and were falling even farther below those market rates as 1979 progressed. Though rates on short-term certificates of deposit (CDs) at thrifts were somewhat higher than their passbook rates, these CD rates were usually still substantially below Treasury bill rates. With differentials of these magnitudes appearing and widening, depositors began actively to search out alternative investments where they could receive a market return on their money. Treasury bills (for those with at least $10,000) were one attractive possibility. Real estate, art, stamps, coins, and other "collectibles" also offered attractive returns, albeit less secure than the interest on an insured deposit.

Most important, however, was the development and growth of money market mutual funds (MMMFs) in the 1970s. These funds pooled the moneys of investors and invested those funds in Treasury bills, in the short-term debt (commercial paper) of high-quality corporations, and in the large denomination (above $100,000) short-term CDs of commercial banks (and a few thrifts). These large denomination CDs were not subject to the Regulation Q ceilings, and thus the MMMF investors could earn a market return on their investments. Though MMMF liabilities were not federally insured, the assets of the MMMFs were generally of high enough quality so that investors' moneys were relatively secure; in addition, the MMMFs usually offered rapid liquidity and limited check-writing privileges.

Table 5-1
Interest Rates on Three-Month Treasury Bills,
1978–1983

	1978	1979	1980	1981	1982	1983
January	6.49%	9.35%	12.04%	14.72%	12.41%	7.81%
February	6.46	9.26	12.81	14.90	13.78	8.13
March	6.32	9.46	15.53	13.48	12.49	8.30
April	6.31	9.49	14.00	13.64	12.82	8.25
May	6.43	9.58	9.15	16.30	12.15	8.19
June	6.71	9.04	7.00	14.56	12.11	8.82
July	7.07	9.26	8.13	14.70	11.91	9.12
August	7.04	9.45	9.26	15.61	9.01	9.39
September	7.84	10.18	10.32	14.95	8.20	9.05
October	8.13	11.47	11.58	13.87	7.75	8.71
November	8.79	11.87	13.89	11.27	8.04	8.71
December	9.12	12.07	15.66	10.93	8.01	8.96

Source: CEA data.

They were clearly an adequate substitute for deposits in the minds of many thrift depositors.

MMMFs had not existed before 1972 and had grown only slowly prior to 1978. As Table 5-2 indicates, however, their assets grew rapidly between 1978 and 1982.

With the rising MMMFs rapidly sucking deposits out of thrifts, Regulation Q was now at best an irrelevance and at worst a cause of disintermediation. Thrifts could try to prevent deposit withdrawals by paying higher interest rates. Indeed, this was tried in June 1978, when the Federal Reserve and the Bank Board loosened Regulation Q slightly to allow banks and thrifts to pay market rates on six-month

Table 5-2
Assets of Money Market Mutual Funds,
1972–1989

Year	Assets[a] (billions)
1972	$0.0
1973	0.1
1974	1.9
1975	3.1
1976	3.0
1977	3.3
1978	9.5
1979	42.9
1980	76.8
1981	188.6
1982	236.3
1983	181.4
1984	230.2
1985	241.0
1986	292.4
1987	310.7
1988	327.0
1989	411.9

[a] Daily average for December of each year.

Source: CEA data.

CDs in denominations above $10,000. (The average deposit in a thrift in 1977 had been approximately $4,750.) Thrifts, however, would then suffer operating losses, since the interest income from their mortgage portfolios would be insufficient to cover their interest costs. Or they could refuse to pay the higher interest rates, watch their deposits depart, and be forced to liquidate their low-interest mortgages at a loss in this high-interest environment. Either route meant losses. There was no way to avoid red ink.[1]

Table 5-3 offers the standard presentation of the industry's profits and losses during this period. It shows the rising losses of the industry in 1979 and the early 1980s. By early 1980, one-third of the industry

Table 5-3
Profits or Losses by FSLIC-Insured Thrifts, 1978–1983

		Unprofitable Thrifts as a Percentage of All Thrifts	Unprofitable Thrifts' Assets as a Percentage of All Thrift Assets	Losses of Unprofitable Thrifts (millions)	Profits or Losses of All Thrifts (millions)
1978:	H1	2.9%	1.5%	$ −25	$1,869
	H2	2.7	1.3	−16	2,051
1979:	H1	4.7	2.6	−35	1,821
	H2	6.5	4.1	−50	1,792
1980:	H1	30.5	30.0	−335	478
	H2	35.5	33.0	−443	303
1981:	H1	69.1	74.8	−1,732	−1,506
	H2	84.8	91.3	−3,324	−3,125
1982:	H1	83.2	89.0	−3,390	−3,205
	H2	67.8	60.6	−2,085	−937
1983:	H1	38.4	33.0	−868	1,101
	H2	35.2	33.2	−1,021	843

Source: FHLBB data.

was reporting losses, and overall industry profitability was greatly reduced. By the first half of 1981 the industry was hemorrhaging and would continue to bleed massively for the next two years.

Though the figures shown in Table 5-3 painted a bleak picture of the financial condition of the thrift industry, in fact they seriously *understated* the losses that a more conservative accounting approach would have presented for the thrift industry during 1981 and 1982. As is demonstrated in the appendix to this chapter, a proper recognition of the writedowns and capital losses from the sales of assets during these years would have shown industry losses, indicated in the middle column of Table 5-4, of $6.9 billion in 1981 (rather than the $4.6 billion shown in Table 5-3) and $22.2 billion in 1982 (rather than $4.2 billion). Thus, from the second half of 1981 through the end of 1982 the thrift industry's hemorrhaging was considerably worse than the usual presentation of the consolidated industry data indicated. Both net operating income and true net nonoperating income were deeply in the red. Further, none of these accounting measures showed the even larger declines in market values that were occurring for the mortgages that were still held by the industry.

Table 5-4
Net Operating Income and Changes in Tangible Net
Worth for All FSLIC-Insured Thrifts, 1981 and 1982

		Net Operating Income (billions)	Change in Tangible Net Worth[a] (billions)	Implied Net Nonoperating Income[a] (billions)
1981:	H1	$-2.2	$-1.6	$+0.6
	H2	-4.9	-5.3	-0.4
			6-9	
1982:	H1	-4.8	-10.6	-5.8
	H2	-3.9	-11.6	-7.7
			22.2	

[a] Includes the effects of $1.5 billion in tax refunds in 1981 and $1.6 billion in 1982; also includes the effects of $145 million paid out in dividends in 1981 and $130 million in 1982 and $127 million capital inflow from mutual conversions in 1981 and $123 million in 1982.

Source: FHLBB data.

By any financial measuring rod, the industry was sliding downhill rapidly.

The Policy Responses

With the thrift industry's losses mounting during 1980-1982, the pressures for a new legislative solution by the Congress (and by individual state legislatures) were enormous. Regulation Q was no longer the answer. Since the creation and maintenance of the thrift industry as a specialized lender had been largely the responsibility of Congress—and since it was the Congress that had extended Regulation Q to the thrifts in 1966 and had discouraged the development of ARMs in the 1970s—the Congress again perceived its responsibility for a new solution.

By the early 1980s the Carter and Reagan administrations, the Congress, and a number of state legislatures had developed a new view of thrifts and their problems. The thrifts' narrow specialization on fixed-interest-rate mortgages was now seen as the major cause of their difficulties, and economic deregulation—a substantial expansion of their asset and liability powers (or, equivalently, a substantial loosening of restrictions)—was seen as the solution. This belief in deregulation was consistent with the deregulation legislation that the Carter administration had proposed and that the Congress had already passed or would shortly pass for the airlines, trucking, and railroad industries.

The first important "action" taken by the Congress was one of nonaction. During 1979 the Bank Board developed and refined a set of regulations that allowed federally chartered thrifts to offer ARMs. Earlier in the 1970s the Congress had shown its hostility to ARMs; now, the Congress failed to raise significant objections.

Next, in early 1980 the Congress passed (and President Carter signed) the Depository Institutions Deregulation and Monetary Control Act (DIDMCA). This Act, though, provided only a beginning. As interest rates continued to remain high through mid-1982 and thrifts continued to hemorrhage, the Congress realized that more action was necessary. The consequence was the Garn–St Germain Depository Institutions Act, signed by President Reagan in late 1982. And during this same time period a number of states provided substantial asset-powers deregulation for their state-chartered thrifts.[2]

The changed economic regulatory environment for thrifts brought about by these actions is best summarized by grouping the changes into those that applied to asset powers and those that applied to liability powers:[3]

Assets. First, thrifts were allowed to offer ARMs. The DIDMCA endorsed and encouraged the Bank Board's regulatory development of ARMs. The Garn–St Germain Act explicitly authorized ARMs for state-chartered thrifts as well.

Second, the DIDMCA and the Garn–St Germain Act expanded considerably the types of loans and investments that federally chartered thrifts could make.[4] Included in this expansion were the ability to issue credit cards and offer consumer loans (e.g., auto loans, personal loans), up to a maximum of 30 percent of a thrift's assets; to make commercial real estate loans, up to a maximum of 40 percent of a thrift's assets; to make commercial loans (secured or unsecured), up to 11 percent of a thrift's assets;[5] and to take direct ownership (equity) positions in ventures, if done through a "service corporation" subsidiary, up to a maximum of 3 percent of a thrift's assets.[6]

Third, individual states provided similar or even wider asset powers for their state-chartered thrifts. Especially important was the ability of state-chartered thrifts in states such as Texas, Florida, and California to invest even higher percentages of their assets in commercial real estate loans, commercial loans, and direct equity positions.

To some extent these state actions were a direct response to the perceived financial difficulties of their state-chartered thrifts. But, to some extent also, the state actions were efforts to induce state-chartered thrifts to refrain from switching to federal charters. As the DIDMCA and the Garn–St Germain Act made federal charters relatively more attractive to thrifts, many states feared that their state-chartered institutions would switch charters, thereby reducing the states' regulatory roles, personnel, and fee collections.[7] There was thus a conscious competition of the states vis-à-vis the federal government to keep and attract thrifts to state charters. The states, though, would bear little or none of the eventual costs of this competition, since the deposit insurance for these state-chartered thrifts was provided by the FSLIC.[8]

These wider asset powers granted by the federal and state governments were not expected to cause thrifts to abandon their role as residential mortgage lenders. Rather, these opportunities to diversify into other types of loans and investments were seen as ways to strengthen thrifts' profitability and (along with ARMs) reduce their susceptibility to interest rate squeezes, since the new types of loans had shorter maturities than the standard thirty-year home mortgage. Further, the expertise and customer base of most thrifts still lay in the area of residential lending; they were unlikely to abandon that base. And an important provision of the tax code, which offered substantial tax reductions to a thrift that had a high percentage of its assets in housing-related investments, remained unchanged.

Liabilities. First, the DIDMCA ordered the phasing-out of the Regulation Q interest rate ceilings for thrifts and banks (except on commercial checking accounts); the Garn–St Germain Act accelerated the phase-out. Though the Regulation Q ceilings were not formally abolished until March 1986, a significant fraction of bank and thrift deposit liabilities were freed from interest rate restrictions by early 1983.

Second, the DIDMCA authorized thrifts and banks to offer interest-paying checking accounts (negotiable order of withdrawal, or NOW accounts) to individuals nationwide.[9] The Garn–St Germain Act authorized thrifts to offer commercial checking accounts to businesses with which they had a lending relationship. Both actions were important in opening up new ways for thrifts to attract funds.

Finally, the DIDMCA increased the maximum insured deposit amount for all federally insured banks, thrifts, and credit unions. In 1974, when the Congress raised the basic insured deposit amount to $40,000 (from the previous $20,000 level), it extended insurance coverage to $100,000 for state and local government deposits. In 1978, the Congress expanded the $100,000 coverage to include deposits by individuals (or their agents) as part of tax-favored retirement plans, such as Keogh plans and individual retirement accounts (IRAs). Now, in 1980 the DIDMCA extended the $100,000 insured amount coverage to all types of accounts.

A Hindsight Analysis and Critique

The major deregulation actions just described were fundamentally sound and sensible. A number of commissions and studies during the previous decade had called for wider powers for thrifts and for the end of Regulation Q.[10] The authorization to offer ARMs was essential for thrifts to deal with their interest rate risk problems, and the opportunity to diversify their asset portfolios, if done prudently, could strengthen their profitability and reduce their overall riskiness. The Regulation Q ceilings had meant distortions and inefficiencies; the higher were the levels of market interest rates, the greater were the distortions and inefficiencies. And with the rise of the money market mutual funds, the Regulation Q ceilings also meant disintermediation and the ultimate nonviability of the ceilings themselves. With the abolition of those ceilings, banks and thrifts could compete freely among themselves (and with the MMMFs) on price, and depositors could receive the full benefit of that competition.

The increase in the insured amount to $100,000 is today a more controversial issue; but this too was a sensible step, since a larger insured deposit amount lessens the vulnerability of a depository to runs by nervous depositors. (A more thorough discussion of this issue will be offered in Chapter 11.) In any event, the $100,000 limit had already been present for some types of accounts, and the general expansion to $100,000 attracted little or no adverse comment at the time.

An important proviso should immediately be added to this characterization of these economic deregulation actions as sensible: *They needed to be accompanied by a substantially stepped-up effort at safety-and-soundness regulation (including better information, tighter scrutiny, and higher net worth standards) and/or the expanded use of economic incentives (e.g., the use of risk-based insurance premiums)*. In the absence of this stepped-up safety-and-soundness regulation too many thrifts were likely to engage in excessive risk-taking—with disastrous consequences for the FSLIC. This risk-taking could (and would) encompass a wide range of behaviors that might lead to higher profits for a thrift but could also lead to larger losses: investing in high-risk assets; plunging into new categories of investments without adequate knowledge; taking less care in the underwriting of new loans; buying fractional participations in real estate loans and projects originated by others without adequate research as to the viability of the project; exercising insufficient caution and placing too much confidence in the continued economic expansion of the oilbelt economy; stretching the limits of regulatory procedures; and engaging in outright fraud and violations of laws and regulations.

To see why a large-scale increase in risk-taking of all kinds was so likely, it is useful to recall the simple balance sheet diagrams of Chapter 3 and reconsider the federal and state actions, as well as the general economic environment of the thrift industry in the early 1980s, in terms of the *opportunities, capabilities,* and *incentives* for taking risks.

Opportunities. The expanded risk-taking opportunities were primarily on the asset side of thrifts' balance sheets. Prior to 1980 thrifts were largely restricted to making home mortgage loans. Though these mortgages created substantial interest rate risk, they usually carried little default or credit risk. Except in the most dire of economic circumstances, mortgage borrowers were generally reluctant to default on their home mortgages. Thrifts strengthened this pattern by becoming adept at weeding out high-risk borrowers. And the regulatory requirement that a borrower should make at least a 10 percent down payment or obtain mortgage insurance further protected thrifts. Defaults were rare. And, in the rising real estate markets of the 1950s, 1960s, and 1970s, thrift losses on defaulted mortgages were rarer still.

The expanded asset powers now gave thrifts the opportunity to enter riskier fields. The potential gains were higher—but so were the potential losses. Though these opportunities could be employed for a prudent diversification strategy, they could also be employed for a deliberate strategy of expanded risk-taking.

Further, the safety-and-soundness regulatory system for thrifts that was in place in the early 1980s was geared to the safe world of the home mortgage. Regulators were unfamiliar with the new types of assets that were now open to thrifts and were unaccustomed to dealing with, or even thinking about, thrifts that were aggressive and risk-embracing; as of 1980 the overwhelming majority of thrifts were still small, mutual, and conservative.

This regulatory system, which was adequate for the sleepy and safe world of the home mortgage, was inadequate for the world of expanded opportunities of the early 1980s. In the absence of a substantial strengthening and reorientation, it would be overwhelmed.

Capabilities. The increased capabilities for risk-taking were primarily on the liabilities side of thrifts' balance sheets. The elimination of the Regulation Q ceilings and the increase in the insured deposit amount meant that thrifts had a vastly increased capability for gathering the funds (deposit liabilities) to finance the expanded opportunities for risk taking.

Prior to 1980 the Regulation Q ceilings constrained thrifts' ability to expand their deposits rapidly. Noncash prizes, such as toasters and tennis rackets, might induce a few depositors to shift their accounts from one institution in town to another down the block; but prizes were unlikely to provide the means for a thrift to expand its deposits rapidly.

With the removal of the Regulation Q ceilings, however, a thrift could now pay market rates of interest—or even slightly above market rates—to attract depositors. Small thrifts in out-of-the-way places could place ads in regional or national financial publications, promising attractive interest rates on *federally insured* deposits. Or they could employ "deposit brokers," including many prestigious Wall Street securities firms, to gather the deposits for them. Further, the increase in the insured amount meant that these deposits could be collected in appreciably larger packages, reducing the transactions costs per dollar of deposit.

Incentives. The expanded incentives for risk-taking could be found on the lower right-hand side of thrifts' balance sheets: their diminished net worths. The losses of the 1980-1982 period, documented in Tables 5-3 and 5-4, greatly eroded the book value net worths of virtually all thrifts.

Tables 5-5, 5-6, and 5-7, using the tangible net worth[11] measure employed in Table 5-4, shows this erosion. As can be seen in Table 5-5, the overall industry suffered a shocking decline in its tangible net worth between the first half of 1981 and the end of 1982. Even those thrifts that remained tangible solvent, shown in Table 5-6, saw their aggregate net worth decline by almost half. And the ranks of the tangible insolvent (Table 5-7) grew to 415 by the end of 1982 and accounted for almost a third of the industry's assets. Also, by 1982 the insolvent thrifts were, on average, the larger institutions in the industry (as is indicated by the last two columns of Table 5-7, where the percentage of industry assets residing in insolvent thrifts was more than double the percentage of the industry's institutions that were insolvent).

These net worth figures, though, reflected only the standard accounting convention of calculating thrifts' asset values at their historical book values[12]; they ignored the decreases in the *market values* of those assets that had occurred in the high interest rate environment of 1981 and 1982. Estimates of the market value net worth of the overall thrift industry in 1981 and 1982 have placed it at a *negative* $100 billion or worse.[13]

With their net worths badly eroded or gone and with losses continuing to mount, the incentives for hundreds of thrifts to embark upon a strategy of deliberate (or even unconscious) risk-taking were greatly increased. The owners of stock thrifts had little or nothing to lose. If they did not do something to reverse the fortunes of their thrifts, it was likely that the Bank Board would appoint a receiver or merge them out of existence. Even the managers of mutual thrifts—

which constituted 80 percent of all FSLIC-insured thrifts in 1980—though they might not have a direct ownership interest to protect, had their positions, salaries, and perquisites at stake and could gain by taking risks. In other instances, new entrepreneurs entered the industry, either buying thrifts that were barely solvent or entering de novo, and were able to expand rapidly (investing in new and risky assets) with little of their own money at stake. Further, they had little in the way of community reputation or community franchise value at stake either.

This opportunities-capabilities-incentives framework is also useful for understanding why the commercial banking sector did not

Table 5-5
Tangible Net Worth for All FSLIC-Insured Thrifts,
1978–1983

		Number of Thrifts	Assets (billions)	Tangible Net Worth (billions)	Tangible Net Worth as a Percentage of Assets
1978:	H1	4,051	$465.0	$26.1	5.6%
	H2	4,048	497.3	28.0	5.6
1979:	H1	4,040	529.1	29.7	5.6
	H2	4,038	554.4	31.5	5.7
1980:	H1	4,021	572.6	31.9	5.6
	H2	3,993	603.8	32.2	5.3
1981:	H1	3,916	622.8	30.6	4.9
	H2	3,751	639.8	25.3	4.0
1982:	H1	3,533	658.0	14.7	2.2
	H2	3,287	686.2	3.7	0.5
1983:	H1	3,206	752.8	4.4	0.6
	H2	3,146	813.8	3.5	0.4

Source: FHLBB data.

experience the same wave of risky expansion as did the thrift industry. First, the banks' opportunities were not greatly expanded. The legislative acts of the early 1980s did not expand commercial bank asset powers significantly. Commercial banks had long had many of the asset powers newly acquired by thrifts in the early 1980s. Banks were familiar with these types of loans, as were bank regulators.

Second, the elimination of Regulation Q and the increase in the insured deposit amount affected banks less than it did thrifts. Banks were much more dependent on commercial checking account deposits, which were exempted from the removal of the Regulation Q ceilings. Banks were also more dependent than were thrifts on large

Table 5-6
Tangible Net Worth for Tangible-Solvent FSLIC-
Insured Thrifts, 1978–1983

		Number of Thrifts	Assets (billions)	Tangible Net Worth (billions)	Tangible Net Worth as a Percentage of Assets
1978:	H1	4,015	$464.8	$26.1	5.6%
	H2	4,010	496.9	28.0	5.6
1979:	H1	4,000	528.8	29.7	5.6
	H2	4,004	554.4	31.5	5.7
1980:	H1	3,980	572.2	31.9	5.6
	H2	3,950	603.4	32.2	5.3
1981:	H1	3,864	621.4	30.6	4.9
	H2	3,641	615.1	26.3	4.2
1982:	H1	3,268	547.7	19.8	3.6
	H2	2,872	466.2	16.4	3.5
1983:	H1	2,741	522.4	17.7	3.4
	H2	2,631	529.2	18.2	3.4

Source: FHLBB data.

denomination CDs, which had been exempted from the Regulation Q ceilings in the early 1970s, and on commercial deposit amounts in excess of $100,000, which continued to be uninsured.

Third, and perhaps most important, banks' incentives were not significantly altered. Banks were not financially devastated by the high interest rate environment of the early 1980s in the way that thrifts were. Bank loans tended to be much shorter in maturity than thirty-year fixed-rate thrift mortgages and frequently carried floating interest rates. This absence of devastation can be seen by comparing the banking industry's profit and net worth results with those of the

Table 5-7
Tangible Net Worth for Tangible-Insolvent
FSLIC-Insured Thrifts, 1978–1983

		Number of Thrifts	Assets (billions)	Tangible Net Worth (billions)	Tangible Net Worth as a Percentage of Assets	Number of Thrifts as a Percentage of All Thrifts	Assets as a Percentage of All Assets
1978:	H1	36	$0.2	$-0.003	-1.4%	0.9%	0.04%
	H2	38	0.4	-0.004	-1.1	0.9	0.04
1979:	H1	40	0.3	-0.007	-2.2	1.0	0.05
	H2	34	0.3	-0.004	-1.6	0.8	0.05
1980:	H1	41	0.3	-0.005	-1.6	1.0	0.05
	H2	43	0.4	0.005	-1.4	1.1	0.07
1981:	H1	52	1.5	-0.030	-2.2	1.3	0.2
	H2	110	24.7	-0.8	-3.2	2.9	3.9
1982:	H1	265	110.3	-5.1	-4.6	7.5	16.7
	H2	415	220.0	-12.8	-5.8	12.6	32.1
1983:	H1	465	230.4	-13.3	-5.8	14.5	30.6
	H2	515	284.6	-14.7	-5.2	16.3	35.0

Source: FHLBB data.

thrifts, as is done in Table 5-8. While thrifts ran large losses in 1981 and 1982 and suffered a major erosion in net worth, bank profits dipped only modestly, and their aggregate net worth remained basically unchanged. Thus, few banks were in the position that characterized many thrifts: "We have little or nothing to lose."

In sum, the legislative acts of 1980-1982 and the economic environment of the thrift industry at this time greatly enhanced the opportunities for thrifts to take risks, their capabilities for funding those risk-taking opportunities, and their incentives for embarking on this strategy. In the absence of corrective regulatory action, trouble was highly likely. As will be shown in the next sections, not only were the corrective regulatory actions neglected, but a number of federal actions actually made the problem worse! In retrospect, it is remarkable that so few thrifts—numbering "only" in the hundreds—embarked on the risk-taking strategy that would lead to their demise (and that of the Bank Board and the FSLIC) and that most of the industry refrained.

Table 5-8
Net Income and Net Worth for Commercial Banks and
FSLIC-Insured Thrifts, 1977–1983

	Commercial Banks		FSLIC-Insured Thrifts	
	Net Income as a Percentage of Assets[a]	Net Worth as a Percentage of Assets[a]	Net Income as a Percentage of Assets	Net Worth[b] as a Percentage of Assets
1977	0.71%	6.9%	0.71%	5.5%
1978	0.77	6.8	0.78	5.6
1979	0.81	6.9	0.65	5.7
1980	0.80	7.0	0.13	5.3
1981	0.77	7.0	−0.72	4.0
1982	0.67	6.9	−0.60	0.5
1983	0.64	6.9	0.23	0.4

[a] Based on domestic assets and liabilities

[b] Tangible net worth.

Sources: FDIC and FHLBB data.

Making it Worse: Lowering Net Worth Standards and Changing the Accounting Rules

It is clear that the conjunction of expanded opportunities, capabilities, and incentives for risk-taking posed substantial dangers for the FSLIC if the safety-and-soundness regulatory system were not immediately strengthened. Unfortunately, not only were these corrective actions not taken immediately, but a number of perverse federal actions exacerbated the problem substantially.

One important set of unfortunate actions involved the reported net worths of thrifts and the net worth regulatory standards that they were expected to meet. As Chapter 3 indicated, net worth is a crucial financial protection for the deposit insurer; it acts like a deductible. The reported net worths of thrifts were sliding rapidly downward in the early 1980s, and their market value net worths were plummeting even faster. The protection for the FSLIC was fast eroding. Unfortunately, during the early 1980s, the Congress, the late Carter administration, the early Reagan administration, and their Bank Board appointees all acted to mask that decline in protection and also to weaken the Bank Board's net worth regulatory requirements themselves.

The first action involved the *lowering* of the Bank Board's net worth standards. Prior to 1980 the minimum net worth requirement for thrifts had been 5 percent of liabilities. The DIDMCA replaced that requirement with a provision that specified only that the minimum net worth for thrifts should be within a range of 3 percent to 6 percent of liabilities, with the Bank Board to set the specific level by regulation. It was clear that the Congress intended that level to be lower. In 1980 the Bank Board lowered the net worth requirement to 4 percent of liabilities, and in 1982 the Bank Board lowered the requirement again, to 3 percent.

A thrift that failed to maintain its net worth level above the minimum requirement would become subject, at least in principle, to tighter regulatory scrutiny and control. By lowering the net worth requirement the Bank Board was reducing its ability to restrain thrifts whose incentives for risk-taking were increasing.

The net worth standards were not adequate even at the higher levels because they were keyed to accounting concepts based on historical book values rather than current market values. Further, the net worth standards involved five-year averages of net worth and of liabilities and, for de novo institutions, a twenty-year phase-in period before the thrift had to comply fully with the standard.[14] Thus, for a thrift that grew rapidly in short bursts, and especially for a growing de novo thrift, the actual net worth needed at the margin to support that growth was only a fraction of the nominal net worth requirement.[15] The lowering of the net worth standard made an inadequate requirement even less effective.

Second, the Garn–St Germain Act authorized the Bank Board to issue promissory notes to foundering thrifts in return for "net worth certificates" from the thrift. The thrift could count the note from the FSLIC as an asset and the certificate as part of its net worth. This note-and-certificate exchange meant that the Bank Board was allowing thrifts to raise their reported net worth levels in a way that offered no added protection for the FSLIC. This provision of the Garn–St Germain Act effectively blessed and expanded a program of note-and-certificate exchanges that the Bank Board already had undertaken.

Third, the Bank Board modified its accounting rules to allow thrifts in some instances to report higher asset values (and have higher net worths) than standard (generally accepted accounting principles, or GAAP) accounting rules allowed. These modified accounting rules, along with the net worth enhancement of the note-and-certificate exchanges (which also contravened GAAP), came to be known as regulatory accounting principles (RAP). What was already an inadequate information system (GAAP), because it largely ignored market values, became even less adequate.

A major means of allowing thrifts to report higher asset values was the practice of "deferred loan losses." Thrifts were reluctant to sell "underwater" assets (i.e., assets with current market values that were below their amortized historical costs), because GAAP required the recognition of the loss at the time of sale, whereas the continued holding of the asset would not require the recognition of any loss (despite the current market value). To induce thrifts to sell these assets and to restructure their portfolios with higher earning assets, the Bank Board allowed thrifts to spread (defer) the recognition of the loss over a period of ten years. In the interim, the unamortized portion of the loss was carried as an asset, even though that "asset" had absolutely no market value whatsoever.

This practice, however, was not as strange as it might seem at first glance. To the extent that GAAP allowed thrifts (and banks) to carry assets at historical cost rather than at current market value, the thrift was overstating the true value of its underwater assets in a similar way. The major difference was that a decline in interest rates would increase the market values of underwater assets still held by the thrift and bring them closer to their accounting values, whereas the deferred-loan-loss "asset" would always have zero market value.

RAP was not all bad. It did allow thrifts a one-time increase in the value of their premises, to reflect current market values ("appraised equity capital"). And it allowed the subordinated debt that thrifts had issued to be included as part of their regulatory requirement for net worth. On balance, however, RAP was definitely a step backward.

In the early 1980s the Bank Board did (surprisingly) briefly flirt with the idea of adopting a market value accounting framework for thrift reporting. A study group within the agency was appointed and duly prepared a report that outlined the advantages and disadvantages to a market value approach.[16] Not too surprisingly, however, the Bank Board shied away from embracing the approach and continued down the opposite path.

Fourth, the Bank Board actively encouraged solvent thrifts to acquire insolvent thrifts. Though many of these transactions required the expenditure of FSLIC cash to fill the negative net worth "hole" of the insolvent thrift,[17] in other instances the Bank Board encouraged the acquiring thrift to create a goodwill asset to cover the difference between the value of the insolvent thrift's (inadequate) assets and its liabilities. Though goodwill can be, and was in many instances, a legitimate recognition of an enterprise's "going concern" or "franchise" value (and is recognized as a legitimate asset by GAAP[18]), in other instances it involved the creation of an asset that did not have real value as protection for the FSLIC. This liberal creation of goodwill assets was yet another weakening of the residual net worth calculation as an indicator of the protection available to the FSLIC.

It is important to realize that these accounting modifications that enhanced reported asset values and hence elevated reported net worths were worse than just a further muddying of an already murky and badly distorted information system; they also weakened the FSLIC's ability to take corrective actions vis-à-vis errant thrifts. Though the Bank Board might have been able to develop a separate set of accounts that would register a more accurate representation of a thrift's net worth, its legal ability to restrain a thrift's actions was largely linked to that thrift's *regulatory* net worth (i.e., the thrift's RAP net worth as established by the Bank Board.) So long as the thrift's

RAP net worth was above the regulatory minimum, it was largely immune from restraining action by the FSLIC; as long as it was showing solvency as measured by RAP, it was largely immune from the appointment of a conservator or receiver—the ultimate power of the FSLIC to control a thrift's actions. The major grounds for the ability of the Bank Board to restrict a thrift's activities or, at the limit, to appoint a conservator or receiver were the thrift's net worth level, its operation in an unsafe and unsound condition, or its dissipation of assets. Because the latter two conditions were more subjective than the first and had never been defined by regulation, the Bank Board almost always relied on inadequate net worth levels or insolvency as its primary grounds for action. Though a sympathetic thrift management might accede to regulatory restraints even though its RAP net worth might have justified resistance, an obstreperous management could legally hide behind its RAP net worth to gain more time for itself and its actions—at a greater eventual cost to the FSLIC.

In taking all of these actions, the political system—the Congress, two presidential administrations, and their regulatory appointees—was responding to the pressures of the thrift industry, who were saying, in essence, "Look, we're good people. We have never caused any trouble. It is intolerable that a large fraction of our members should be perceived by the public to be in violation of the Bank Board's net worth regulatory standards. We were not responsible for this high interest rate environment that has eroded our net worths, and, besides, we had no choice but to make thirty-year fixed-interest-rate loans. Give us time. Interest rates will surely decline, and the industry can surely heal itself." Unfortunately, this argument ignored the new environment—the changed opportunities, capabilities, and incentives—that threatened to cause much more harm to the FSLIC. By heeding this call for leniency, the political system greatly weakened the FSLIC's ability to protect itself against the much greater dangers to which it was now exposed.

The immediate effects of these changes in accounting rules and the lowering of the net worth standards can be readily documented. First, they staved off the reporting of insolvency by hundreds of thrifts. In Table 5-9 the first two columns show the numbers and assets of firms that were reporting insolvency on the basis of the RAP accounting standards established by the Bank Board. These numbers were quite modest. The next two columns show the number and assets of thrifts that showed *solvency* on the basis of RAP but would have reported *insolvency* on the basis of GAAP; this was the group that RAP was directly saving from otherwise having to report insolvency.[19] This group grew rapidly, and by 1984 it constituted over 10 percent of the industry (with slightly less than 10 percent of the industry's assets). The final two columns show the numbers and assets of thrifts that showed solvency on the basis of GAAP only because of the presence of their "intangible" goodwill assets and that would have been insolvent in the absence of goodwill; their tangible assets were inadequate to cover their liabilities, and thus they were tangible insolvent. The Bank Board's liberal encouragement of the creation of these goodwill assets was an important cause of the rapid growth of this group. By 1984 they had over a quarter of the industry's assets.

Thus, by 1984 the combination of RAP and a liberal amount of goodwill assets was allowing a fifth of the industry, with over a third of its assets, to avoid reporting insolvency.

A second effect was to buoy up the reported net worth (and hence the appearance of financial health) of the industry generally. As Table 5-10 indicates, the ratios of reported RAP net worth to assets for the aggregate industry substantially exceeded the ratios that would

Table 5-9
Numbers and Assets of Insolvent Thrifts, 1981–1984

	RAP-Insolvent Thrifts[a]		RAP-Solvent but GAAP-Insolvent Thrifts[b]		GAAP-Solvent but Tangible-Insolvent Thrifts[a]	
	Number	Assets (billions)	Number	Assets (billions)	Number	Assets (billions)
1981	33	$3.3	54	$10.4	23	$10.9
1982	71	12.8	166	51.5	178	155.7
1983	48	12.5	245	66.5	222	205.5
1984	71	14.9	374	95.4	242	248.0

[a] Thrifts that were reporting insolvency on the basis of regulatory accounting principles (RAP).
[b] Thrifts that were reporting solvency on the basis of RAP but would have been insolvent on the basis of generally accepted accounting principles (GAAP).
[c] Thrifts that would have been solvent on the basis of GAAP only because of the presence of "intangible" goodwill assets and hence would have been insolvent if only "tangible" assets were counted.

Source: FHLBB data.

Table 5-10
Net Worth as a Percentage of Assets, FSLIC-Insured
Thrifts, 1981–1984

	RAP Measurement Basis	GAAP Measurement Basis	Tangible Measurement Basis
1981	4.3%	4.2%	3.9%
1982	3.6	2.9	0.5
1983	3.9	3.0	0.4
1984	3.7	2.6	0.3

Source: FHLBB data.

have been reported on a GAAP basis and greatly exceeded their tangible net worth ratios.

Third, the lowering of the net worth standards themselves allowed hundreds of thrifts to avoid being in violation of the standards and thus avoid restrictive or restraining actions by the Bank Board. The thrifts that benefitted were those with RAP net worth below 5 percent of liabilities (the pre-1980 standard) but above 3 percent of liabilities (the 1982 standard).[20] As Table 5-11 indicates, from 1982 through 1984 over 30 percent of the thrift industry, accounting for almost half of the industry's assets, was in this category and thereby escaped the possibility of more stringent regulatory scrutiny or control.

Table 5-11
FSLIC-Insured Thrifts with Ratios of RAP Net Worth to
Liabilities That Were Less Than 5 Percent and at or
above 3 Percent, 1982–1984

	Number	Percentage of Industry	Assets (billions)	Percentage of Industry Assets
1982	1,077	32%	$337	49%
1983	1,104	35	$377	46
1984	1,041	33	$446	46

Source: FHLBB data.

Making It Worse: Reducing Regulatory Scrutiny

There were two other federal actions that made a bad situation even worse. The first was related to the number of the field-force regulatory personnel—the thrift examiners and supervisors. A significant expansion in their number would have been an essential component of any strengthening of the safety-and-soundness regulatory framework. The early 1980s, however, were a period of a general belief in deregulation, and many members of the Washington policy community could not distinguish between *economic* regulation (and deregulation) and *safety* regulation. Deregulation to them meant a general decrease in the presence of government in the economy.[21] In this spirit, the number of the field-force thrift examiners and supervisors were actually *reduced* between 1981 and 1984, as can be seen in Table 5-12. Further, the numbers of examinations (audits) of thrifts were substantially reduced between 1980 and 1983 and rose only slightly in 1984. The Bank Board's justification for this reduction was the reduced number of thrifts in the system. But, as Table 5-13 indicates, the annual numbers of examinations per thrift fell after 1982, and the number of examinations per billion dollars of assets in the industry fell continuously during the early 1980s.[22]

Table 5-12
FHLBB Regulatory Resources, 1979–1984

	Examination and Supervision Staff[a]	Examination and Supervision Budget (millions)
1979	1,282	$41.0
1980	1,308	49.8
1981	1,385	52.8
1982	1,379	57.3
1983	1,368	62.5
1984	1,337	67.0

[a] Full-time equivalents, including nonprofessionals and support staff.

Source: Barth and Bradley (1989).

The second change was more accidental, though it still had serious adverse consequences. In September 1983 the headquarters of the Ninth District of the Federal Home Loan Bank System—covering the states of Arkansas, Louisiana, Mississippi, New Mexico, and Texas—was moved from Little Rock to Dallas. This district office housed the lending personnel for FHLB loans to thrifts in these five states and, more important, the supervisory personnel for these same thrifts. Also, the regional office for the examiners of the thrifts in these five states made the same move.

In the abstract, this move made a great deal of sense. Little Rock was not a regional financial center; Dallas was. The Federal Reserve System had one of its twelve regional banks in Dallas. The Little Rock office had had difficulties in recruiting personnel to work in Little Rock; recruitment would be easier in Dallas.

The move did cause serious administrative disruption, however, and its timing was exquisitely bad. Only eleven of forty-eight supervisory personnel in the Little Rock office made the move to Dallas, so the new office had to restaff itself substantially, and its supervisory capabilities were clearly weakened.[23] Though all but one of the district's examiners made the move, the disruption of the move, combined with the deregulation philosophy just mentioned, caused the number of examinations in these five states to fall sharply. As can be seen in Table 5-14, the number of examinations in the district fell

Table 5-13
FSLIC-Insured Thrift Examinations, 1980–1984

	Examinations	Number of FSLIC-Insured Thrifts	Thrift Industry Assets (billions)	Examinations per Thrift	Examinations per Billion Dollars of Assets
1980	3,210	3,993	$593.8	0.80	5.41
1981	3,171	3,751	639.8	0.85	4.96
1982	2,800	3,287	686.2	0.85	4.08
1983	2,131	3,146	813.8	0.68	2.62
1984[a]	2,347	3,136	976.9	0.75	2.40

[a] Includes special, limited-scope examinations.

Sources: FHLBB data; Barth, Bartholomew, and Bradley (1989).

by almost one-third between the twelve months preceding the move to Dallas and the twelve months following it. The number of examinations did not exceed premove levels until the third year after the move.

Thus, at just the time that the thrift industry's opportunities, capabilities, and incentives for risk-taking were greatly expanded, the degree and extent of regulatory scrutiny and control was diminishing—especially in the Southwest. The managers of thrifts, particularly in the Southwest, surely realized that they were subject to less scrutiny and control, and this realization in turn surely increased their inclinations to take risks of many varieties.

No Cassandras

One remarkable fact about this period and these deregulation actions is that there were no "Cassandras" at the time; there were no "voices in the wilderness" warning about the extreme dangers that could follow from the DIDMCA and the Garn–St Germain Act.

It is true that a few academic writers had recognized that there was

Table 5-14
Thrift Examinations in the Southwest, 1982–1986

| | Before Relocation[a] | After Relocation[a] | | |
	10/82–9/83	10/83–9/84	10/84–9/85	10/85–9/86
Arkansas	15	8	22	42
Louisiana	50	50	41	70
Mississippi	17	24	19	26
New Mexico	13	1	0	8
Texas	166	100	91	137
Total	261	183	173	283

[a]The headquarters of the Ninth District of the Federal Home Loan Bank System, encompassing the five states shown in the table, was relocated from Little Rock to Dallas in September 1983.

Source: Dochow (1989).

a generic problem in the deposit insurance system: All three deposit insurance funds systematically encouraged risk-taking because they charged flat-rate premiums that were insensitive to risk and they did not measure net worth in market value terms.[24] These generic warnings about the problems of the deposit insurance system, accompanied by the recognition that the generic problem would become worse because of the increase in the insured deposit amount to $100,000, continued through 1982 and 1983.[25]

Further, at the end of a generally laudatory twenty-nine-page review of the Garn–St Germain Act in early 1983, a group of economists at the Federal Reserve Bank of Chicago did note the following:

> Here [the Garn–St Germain Act] the intention is to allow thrifts in particular, to diversify their portfolios in order to reduce their (and, ultimately the FSLIC's) exposure to interest rate risk. However, use of these powers may at the same time increase thrift and corporation exposure to default risk. While the balance has been judged in favor of deregulation at this time, that balance may not always be so.
>
> ...some banks have undertaken risky operations in the past, and will again in the future. As deposit insurance premiums do not reflect risk, risk-takers expose the insurance agencies (and ultimately other depository institutions) to loss. In the past, unacceptable degrees of risk-taking have been prevented largely through regulations that preclude unacceptable behavior. As the deregulatory process successfully removes restraints on depository institution behavior, new ways must be found to forestall unacceptable behavior, possibly by pricing insurance according to risk exposure.[26]

In no instances, however, were the words of caution about a distorted deposit insurance system accompanied by strong words of warning that the deregulation measures of the DIDMCA, the Garn–St Germain Act, and the individual states would—unless immediate countervailing measures were taken—cause these problems to become a great deal worse *with respect to thrifts*. No one was writing (or testifying) in 1982 or 1983 that the legislative and regulatory actions of 1980–1982 had created the potential for a sharp increase in risk-taking by thrifts and a sharp increase in the exposure to loss by the FSLIC and that an immediate and massive strengthening of the safety-and-soundness regulatory system was vital.

Instead, virtually everyone within the Washington policy community (and outside it as well) was mesmerized by the hemorrhaging of the thrifts and focused myopically on measures that would stop the bleeding. And, besides, the thrift industry was still considered to be a sleepy, unaggressive industry that had never caused any harm before. The *enhanced* opportunities-capabilities-incentives nexus was simply not seen—except by the entrepreneurs who would take advantage of it.

This silence on the safety risks for the thrift industry is in striking contrast to the policy warnings that had accompanied the economic deregulation of the airline industry only a few years earlier. In that case the advocates of *economic* deregulation frequently emphasized that *safety* regulation of the airlines (by the Federal Aviation Administration) was separate and should not be weakened.[27]

The spectre of airplanes crashing to the ground had an immediacy that evoked caution. The possibility of thrifts crashing financially apparently seemed too remote.

A Further Irony

In March 1983 the Bank Board delivered to the Congress a report on the deposit insurance system, which had been mandated by the Garn–St Germain Act.[28] In the light of the agency's actions during the previous two years, the report is quite remarkable. The report recognized that the FSLIC was exposed to more risk:

> In the future, insurance agencies must limit risk through regulations that constrain the activities of insured institutions or through pricing mechanisms that provide proper incentives for risk-taking. If neither option is available, then the insurance agency is exposed to considerable risk.
>
> Although the deregulation of the past few years was a necessary response to marketplace innovations, it has also substantially limited the ability of regulatory agencies to constrain the risk-taking of insured institutions. Moreover, this has occurred at a time when there are a number of insured institutions that are operating with impaired capital and have strong incentives to engage in very risky investments. In light of the competitive pressures that the industry will face in the next few years, this deregulation could result in substantial losses.[29]

The report endorsed market value accounting, higher net worth levels, and risk-sensitive deposit insurance premiums

Unfortunately, the Bank Board's report was completely at odds with the agency's policies and actions during the previous two years and with the positions that it had adopted publicly. Further, two months after the release of the report, the leadership of the agency changed. The report was quickly shunted aside and attracted little notice.

A Summing-Up of the Perverse Responses

At a time of expanded exposure to risk for the FSLIC insurance fund, the federal government chose to reduce the deductible (thrifts' net worths) in the FSLIC insurance arrangement, weaken the FSLIC's information (accounting) system, relax the rules (regulations), and reduce the resources (examiners and supervisors) devoted to scrutinizing and limiting the behavior of the FSLIC's insureds. The federal government also continued its fifty-year tradition of ignoring the arguments in favor of risk-sensitive deposit insurance premiums.[30]

A private insurance company that attempted such a business strategy would soon find itself bankrupt. The demise of the FSLIC took a little longer.

Appendix to Chapter 5

The chapter stated that the publicly reported losses for the thrift industry for 1981 and 1982 were understatements of the losses that a more conservative accounting approach would have indicated. This appendix substantiates that claim.

An understanding of this point requires an explanation of the profit or loss data in Table 5-3. They are the sum of two categories of gains or losses: net *operating* income, which is the net result of a thrift's current revenues from interest received (on mortgage loans and other investments), less its interest costs (on deposits and other borrowings) and other operating expenses; and net *nonoperating* income, which is the net result of capital gains or losses on asset sales and writedowns of assets that have diminished in value.

The first column of Table 5-4 shows the net operating losses for all thrifts during the first and second halves of 1981 and 1982. As can be seen, the net operating losses were substantially larger than the net overall losses reported in Table 5-3. For all of 1982 the net operating losses were $8.7 billion.

The difference between the net operating losses of Table 5-4 and the net overall losses of Table 5-3 would otherwise seem to imply net nonoperating *gains* (e.g., the sale of assets at a profit.) Such gains were highly unlikely in the environment of high interest rates and diminished asset values of 1981 and 1982. Instead, the liberal creation of goodwill as an asset papered over the losses on underwater assets as they were transferred from insolvent thrifts to acquirers; if some of these assets could be subsequently sold at gains from their transfer prices, these gains would be registered. Also, the special, lenient set of accounting rules (regulatory accounting principles or RAP) developed by the Bank Board for the thrift industry during the 1981–1983 period provided the recognition of some other gains.

The second column of Table 5-4 shows the changes in tangible net worth, which exclude the effects of the Bank Board's special actions. These changes were basically the result of the combined consequences of net operating losses and *actual* (GAAP) net nonoperating losses.[31] As can be seen, tangible net worth declined continuously during these two years—and by more than the net operating losses in the last half of 1981 and during all of 1982. The third column of Table 5-4 shows the implied net nonoperating losses for these years. As would be expected, they were also sizably negative from the second half of 1981 onward.

The changes in tangible net worth are probably the best approximation of the industry's overall (GAAP) losses during these two years. They show losses of $6.9 billion and $22.2 billion for 1981 and 1982, respectively, rather than the $4.6 billion and $4.2 billion that was shown in Table 5-3.

Notes

1. Though thrifts could borrow from their local Federal Home Loan Bank at relatively favorable rates, these rates too were much above the interest yields on the thrifts' portfolios of mortgages.
2. Texas had led the way, providing substantial deregulation for its state-chartered thrifts during the 1960s and 1970s. But, as late as 1978, state-chartered Texas thrifts still had 82 percent of their assets in fixed-rate long-term home mortgages; see Fabritius and Borges (1989, Ch. 4 and p. 77). It was the expanded capabilities and incentives discussed in the next section, along with the opportunities that they already had, that led to the expanded risk-taking by the Texas thrifts.
3. The DIDMCA and the Garn–St Germain Act had many other provisions, not discussed here, that applied to banks as well as thrifts. For a summary of the DIDMCA, see Brewer et al. (1980); for a summary of the Garn–St Germain Act, see Garcia et al. (1983).
4. DIDMCA also authorized thrifts to offer trust and fiduciary services.
5. This authorization included high-yield, below-investment-grade corporate debt securities (usually called "junk bonds") that were not permitted as assets for commercial banks.
6. Again, this authorization for direct ownership positions was one that was not permitted for commercial banks.
7. Some cynical observers suggested that the switching of charters would also reduce political contributions to state legislators.
8. The exceptions were the state-chartered thrifts in Ohio and Maryland that were insured by state-sponsored funds, both of which collapsed in 1985.
9. NOW accounts had first been authorized for state-chartered mutual savings banks in Massachusetts in the early 1970s. They had spread to other thrifts and to commercial banks in New England, New York, and New Jersey by the end of the 1970s.
10. See Jacobs and Phillips (1972), Phillips and Jacobs (1983), and White (1986).
11. "Tangible net worth" measures net worth according to GAAP, except that goodwill assets are excluded from the calculation. Though goodwill can be a legitimate asset, as will be argued later in this chapter, it can also be another way of artificially bolstering asset values and hence net worth. To abstract from this problem of evaluating the goodwill assets

that were created during 1981 and 1982, tangible net worth has been used to represent the changes in the accounting measures of net worth that were occurring at the time.

12. As was noted earlier, the tangible net worth figures ignore goodwill.

13. See Kane (1985, p. 102; 1989, p. 74); Brumbaugh (1988, p. 50); and Brewer (1989, p. 3).

14. The rules related to averaging and the phase-in period preceded the 1980s. They were designed to accommodate mutual thrifts, which could not sell stock to raise capital and were thereby limited to retained earnings (reinvested profits) as the means for increasing net worth.

15. For example, if a thrift had a 3 percent net worth requirement and had liabilities of $100 million without change for five years, its net worth requirement would be $3 million. If, in the sixth year, it grew to $200 million in liabilities, its five-year average of liabilities would be only $120 million, so its net worth requirement would be only $3.6 million. Thus, the thrift would need only $0.6 million of additional net worth in the first year to support the additional $100 million in liabilities, or an effective capital requirement of only 0.6 percent at the margin.

16. See FHLBB (1983b) and Hempel et al. (1989).

17. During 1981 and 1982 the FSLIC liquidated two insolvent thrifts, at an estimated cost of $33 million, and placed 89 insolvent thrifts with acquirers, at an estimated cost of $1.5 billion.

18. It is not generally recognized as an asset by bank regulators.

19. Thrifts that had publicly traded securities were required to report to their securities holders on the basis of GAAP; but their reports to the Bank Board were on the basis of RAP.

20. A thrift could find itself in this range because of losses that reduced its net worth from a higher level *or* because its assets grew while its net worth level remained unchanged.

21. For a description of the difficulties that the Bank Board faced when it finally tried to expand its regulatory resources in late 1984 and early 1985, see Pizzo et al. (1989, pp. 267-269).

22. On average, each examination does appear to have become somewhat more thorough, since the number of examiner working days per examination rose moderately during these years; but the number of examiner working days per billion dollars of assets fell sharply. Also, in theory the slack might have been taken up by the state thrift regulators, since the examinations of state-chartered thrifts were also their responsibility; but this does not seem to have been the case in practice.

23. See Cole (1989a) and *The Washington Post,* June 11–17, 1989.

24. For example, see Scott and Mayer (1971); Karaken and Wallace (1978); Sharpe (1978); Kane (1980); and Buser et al. (1981).

25. For example, see Kane (1982; 1983a; 1983b), Flannery (1982), and Horvitz (1983c).

26. Garcia et al. (1983, pp. 29-30).
27. See Meyer et al. (1981, p. 10).
28. FHLBB (1983a).
29. FHLBB (1983a, pp. 41-42).
30. See the articles cited in Notes 24-26 and FHLBB (1983a, Section IV).
31. Corrected for tax refunds, dividends paid out to shareholders, and new capital inflows.

CHAPTER SIX

The Debacle, 1983–1985

Interest rates declined in the last half of 1982 and into 1983. Thrifts' deposit costs came down, and most of the thrift industry began earning profits again. The industry was no longer in the interest rate vise of the previous four years, and most thrifts remained with their traditional focus on home mortgages—albeit with a greater sprinkling of ARMs and a few new assets.

The new powers described in Chapter 5 were open to all, however, and a fraction of the industry—a minority that nevertheless numbered in the hundreds—wasted little time in taking advantage of the expanded opportunities, capabilities, and incentives that were available to them. Over the next three years—1983-1985—the overall thrift industry grew rapidly. By the end of 1985, the industry was more than half again (56 percent) as large as it had been only three years earlier. Hundreds of individual thrifts grew far faster, doubling or tripling in size over these three years, and some grew by even higher multiples. For these hundreds of thrifts this rapid growth in assets meant new and unfamiliar types of loans and investments, which would subsequently decline in value.

The losses reported by these thrifts, and their eventual insolvencies, did not begin to become significant until after 1985, because of delays engendered by accounting rules and because real estate values in the Southwest (and especially Texas) began to drop precipitously only then. This delay in the recognition of losses until the late 1980s misled many legislators and media reporters (and their audiences) to believe that those losses reflected concurrent misbehavior by thrifts that better concurrent regulation should have been able to prevent. Unfortunately, in most instances the losses were already embedded in the assets on the thrifts' balance sheets and could not be avoided.[1]

Thus, the real debacle occurred largely between 1983 and 1985; but the revelation and recognition of that debacle would only occur later in the decade.

Rapid Growth

Table 6-1 documents the growth rates of the overall thrift industry in 1983 and 1984. Annual growth rates of assets for those years were in double digits and more than double the annual growth rates of the previous three years.[2] The growth spurt in some individual "sunbelt" and energy-related states were considerably sharper and, in some instances, extended into 1985. By the end of 1985 the Texas thrifts (in aggregate) had grown to more than twice their size in the three years since the end of 1982; the same was true of the thrifts (in aggregate) in Arizona.

These growth rates were not, of course, uniform across the entire industry. The growth rates for the individual states of Table 6-1 illustrate this diversity. Another demonstration of this point is provided by an examination of the growth records of the thrifts that were liquidated or placed with acquirers during 1986-1989, or were predicted (as of early 1989) to require disposal. There were 637 soon-to-fail thrifts in this group that had financial reports available for both

Table 6-1
Annual Percentage Growth Rates of FSLIC-Insured
Thrifts,[a] 1980–1986

	1980	1981	1982	1983	1984	1985	1986
Total U.S.	7.2%	7.8%	7.3%	18.6%	19.9%	9.5%	8.7%
Arkansas	6.0	4.2	−2.3	42.9	24.7	6.2	−2.0
Arizona	13.5	9.4	23.5	18.3	46.7	23.8	15.3
California	10.1	8.2	18.3	28.0	29.6	8.8	13.1
Colorado	13.0	8.6	−9.2	9.9	24.7	6.8	12.1
Florida	11.8	10.5	9.6	17.1	20.7	7.6	2.2
Kansas	6.6	5.3	5.9	21.2	28.8	20.9	11.9
Louisiana	10.5	8.0	11.5	20.3	18.1	3.7	5.2
Oklahoma	11.7	10.4	9.8	16.5	13.6	5.4	1.9
Texas	11.9	9.7	13.2	33.3	38.0	18.4	5.5

[a] Growth in assets, from previous year end.

Source: Barth, Bartholomew, and Bradley (1989).

1982 and 1985. These thrifts had $140.8 billion in assets at year-end 1982. Three years later this group had grown by 101 percent, doubling in size to $282.9 billion in assets.[3] By contrast, the remainder of the industry grew by only 49 percent during those three years. Among this group of 637 eventual failures, 74 grew by more than 400 percent between 1982 and 1985 (i.e., by year-end 1985 they were more than five times their sizes of three years earlier). The "champions" among these "flameouts" were Diversified American Savings Bank of Lodi, California, which grew from $11 million in assets in 1982 to $978 million in assets in 1985, and a de novo thrift, Bloomfield Savings and Loan of Birmingham, Michigan, which grew from $2 million in assets in 1982 to $676 million in 1985.

This growth spurt by thrifts after 1982 was partly due to the recovery of the U.S. economy after the recession of the early 1980s, but much more was involved. As can be seen in Table 6-2, thrift growth rates greatly exceeded the growth of nominal gross national product (GNP) and also exceeded the growth in mutual savings bank assets

Table 6-2

Annual Percentage Growth Rates of Various Types
of Depository Institutions[a] and of Nominal GNP,
1980–1986

	1980	1981	1982	1983	1984	1985	1986
FSLIC-insured S&Ls	7.2%	7.8%	7.3%	18.6%	19.9%	9.5%	8.7%
FDIC-insured MSBs	3.7	2.3	0.0	9.9	4.8	14.7	15.4
FDIC-insured commercial banks:							
Domestic assets	9.5	9.8	11.0	8.0	4.0	10.4	8.8
Worldwide assets	9.7	9.3	8.1	6.7	7.1	8.9	7.7
Nominal GNP[b]	9.4	11.0	2.6	10.4	8.6	6.6	4.8

[a] Growth in assets, from previous year end.
[b] Measured by the fourth quarter GNP in each year.

Sources: FHLBB, FDIC, and CEA data.

and commercial bank assets during these same years. The absence of a rapid expansion by the MSBs and commercial banks reinforces the point made in Chapter 5: Banks did not face the same set of expanded opportunities, capabilities, and incentives as did FSLIC-insured thrifts.

Rapid growth by any business enterprise is likely to involve management and organizational problems; thrifts are no exception. Compounding these usual problems was the fact that this growth often involved new categories of loans and investments. Thus, there were new opportunities to take risks—or just to make mistakes.

An indication of the growth in new types of assets in the early 1980s is provided in Table 6-3. As can be seen the aggregate amount of "nontraditional" assets held by thrifts almost tripled between 1982 and 1985; the absolute increase was about $137 billion. By the latter year over a fifth of thrifts' aggregate assets was invested in these nontraditional categories.

This expansion was largely mirrored by the decline in the relative

Table 6-3
Holdings of "Nontraditional" Assets by FSLIC-Insured
Thrifts, 1982 and 1985

	1982		1985		
	Amount (billions)	Percentage of Total Assets	Amount (billions)	Percentage of Total Assets	Increase in Amount, 1982–1985 (billions)
Commercial mortgage loans	$43.9	6.4%	$98.4	9.2%	$54.5
Land loans	6.9	1.0	31.0	2.9	24.1
Commercial loans[a]	0.7	0.1	16.0	1.5	15.2
Consumer loans	19.2	2.8	43.9	4.1	24.7
Direct equity investments	8.2	1.2	26.8	2.5	18.6
Total	$78.9	11.5%	$216.1	20.2%	$137.2

[a] Includes "junk bonds."

Source: Barth, Bartholomew, and Labich (1989).

importance of thrifts' traditional investment: residential mortgages. Though thrifts' holdings of residential mortgages[4] grew by $120 billion between 1982 and 1985, the *percentage* of thrifts' assets that were in residential mortgages declined during these same three years, from 65 percent in 1982 to 53 percent in 1985.[5]

An equally revealing picture of the connection between rapid growth and new types of assets can be found in Table 6-4. The correlation between thrifts that were growing rapidly in 1984 and the percentage of their assets that was devoted to nontraditional investments was very high indeed. Further, the percentage of residential mortgages (and mortgage-backed securities) in their portfolios and their growth rates were negatively correlated.

The funding for the rapid growth tended to come from sources other than the traditional retail or "core" deposit base. These new sources included large denomination deposits (above $100,000) that

Table 6-4
Asset Growth and the Asset Composition of
FSLIC-Insured Thrifts, June 1984

	Thrifts' Annual Rates of Asset Growth[a]			
	Less than 15%	15 to 25%	25 to 50%	More than 50%
Composition of assets:				
Residential mortgages and mortgage-backed securities[b]	68.1%	67.1%	63.4%	53.0%
Commercial mortgages	6.6	8.3	8.2	10.8
Land loans	1.2	1.8	2.5	5.8
Nonmortgage loans	3.7	3.6	4.6	5.0
Real estate investments	0.2	0.6	0.7	1.2
Other assets	20.2	18.6	20.6	24.2
Total	100.0%	100.0%	100.0%	100.0%

[a] Annual rate of growth extrapolated from growth in the first half of 1984.
[b] Includes the category "other mortgage-related assets."

Source: Mahoney and White (1985).

were either gathered with the help of outside deposit brokers or through in-house "money desks" and repurchase agreements ("repos") with investment banking firms.[6] Table 6-5 shows the positive correlation of thrifts' growth rates and their reliance on these nontraditional funding sources.

Though the industry had previously experienced periods of rapid growth during the 1945-1979 era, that growth had been almost entirely in the traditional line of business with which thrifts were familiar and comfortable. Also, much of that expansion was due to inflation and larger mortgages rather than to a higher volume of mortgages, so the management problems associated with a higher *volume* of transactions were less likely to be present. This last point is illustrated in Table 6-6. For example, between 1965 and 1979 total thrift assets and thrift mortgage assets in particular increased by over four-fold. But the total number of mortgage *holders* increased by only three-fifths, and even the number of mortgage holders per thrift

Table 6-5
Asset Growth and the Liability Composition of
FSLIC-Insured Thrifts, June 1984

	Thrifts' Annual Rates of Asset Growth[a]			
	Less than 15%	15 to 25%	25 to 50%	More than 50%
Composition of liabilities:				
Retail deposits	80.9%	72.4%	64.7%	59.0%
Large denomination deposits[b]	7.3	11.6	14.5	18.1
Repurchase agreements	2.4	4.0	6.8	10.4
FHLB advances	6.7	8.1	9.6	7.3
Other liabilities	2.7	3.9	4.4	5.2
Total	100.0%	100.0%	100.0%	100.0%

[a] Annual rate of growth extrapolated from growth in the first half of 1984.
[b] Above $100,000.

Source: Mahoney and White (1985).

increased by only four-fifths; these increases implied annual growth rates of only 3.5 percent and 4.3 percent, respectively. The last column of Table 6-6, which presents the number of new mortgages in each year (i.e., the new transactions activity in that year) per thrift, tells the same story. Thus, the growth experience of the pre-1980 era was quite different from that of the 1983-1985 period.

Another indicator of change and growth in the thrift industry in the 1980s was a wave of new entrepreneurs that entered the industry. Their entry took a number of forms. First, they started new (de novo) thrifts. Table 6-7 shows that 268 thrifts were approved to receive FSLIC deposit insurance during 1983-1985, more than double the 119 that received approval during the previous three years.

A second form involved the conversions of mutual thrifts into stock companies. As can be seen in Table 6-8, the number of completed conversions during the 1983-1985 period—257—was more than triple the number completed during the previous three years. In some instances the former managers of the mutual thrift stayed in control; in others, especially where the former mutual was close to

Table 6-6
FSLIC-Insured Thrifts' Mortgage Assets and Mortgage
Holders, 1950–1979

	Number of Thrifts	Total Assets (billions)	Mortgage Assets (billions)	Number of Mortgage Holders (thousands)	Number of Mortgage Holders per Thrift	Number of New Mortgage Loans Made per Thrift[a]
1950	2,860	$13.7	$11.2	2,695	942	98
1955	3,544	34.2	28.7	4,987	1,407	161
1960	4,098	67.4	56.8	7,157	1,746	156
1965	4,508	124.6	106.3	9,622	2,134	194
1970	4,365	170.6	146.0	10,638	2,437	124
1975	4,078	330.3	272.5	13,095	3,211	276
1979	4,039	568.1	467.3	15,504	3,839	352

[a] Includes non-FSLIC-insured thrifts.

Sources: FHLBB and U.S. League of Savings Institutions data.

failing, new entrepreneurs gained control. And, in a third form of entry, new entrepreneurs bought out the interests of the previous owners of small stock thrifts.[7]

In sum, the overall thrift industry grew rapidly between 1982 and 1985, and the thrifts that would subsequently fail grew yet more rapidly. This growth far exceeded the growth of the U.S. economy or of other types of depository institutions. Much of this growth involved new and nontraditional assets; and new and nontraditional entrepreneurs were often at the helms of these rapidly growing thrifts. Though most thrifts refrained from excessively risky activities, a sizable minority embraced them, with delayed but disastrous consequences.

Compounding the Problem

The economic deregulation of the thrift industry would surely have meant some dislocations and failures in any event. Deregulation meant a new operating environment, which required new skills and talents. Some incumbents would be able to adapt; others would fall by the wayside. This has been the pattern in other deregulated industries,[8] and it surely would have applied to thrifts as well.

Table 6-7
Approvals of De Novo Thrifts to Receive Deposit
Insurance from the FSLIC, 1980–1986[a]

	Federally Chartered Institutions	State-Chartered Institutions	Total
1980	5	63	68
1981	4	21	25
1982	3	23	26
1983	11	36	47
1984	65	68	133
1985	43	45	88
1986	14	13	27

[a] Excludes state-chartered thrifts that converted from state-sponsored insurance funds to the FSLIC.

Source: FHLBB data.

Further, the general economic environment for financial institutions was changing: The relatively high inflation rates of the late 1970s were lowered considerably in the 1980s. The legal and regulatory environment was changing, bringing more financial institutions into competition with each other. And the technologies that lie at the heart of financial services—data processing and telecommunications—were also rapidly changing and improving. All of these changes, again, would have meant a new environment for thrifts and a new winnowing of successful and unsuccessful thrifts.[9] The experience of the commercial banks, which were exposed to the same set of changes in their environment (and were also exposed to the loosening of the Regulation Q ceilings) is instructive. As can be seen in Table 6-9, the failure rates for commercial banks rose sharply in the mid-1980s, as compared with the failure rates earlier in the 1980s or with the failure rates in earlier decades that were shown in Table 4-3.

But the experience of the FSLIC-insured thrifts, which is also shown in Table 6-9, was clearly much more severe. As Chapter 5 demonstrated, the enhanced opportunities, capabilities, and incentives for risk-taking—exacerbated by perverse federal actions that reduced the effectiveness of safety-and-soundness regulation—created an explosive mix. The rapid growth of many thrifts, especially those involved in new types of assets, indicated that they were taking

Table 6-8
Conversions of FSLIC-Insured Mutual
Thrifts to Stock Thrifts, 1980–1986

	Number of Conversions
1980	16
1981	32
1982	31
1983	83
1984	96
1985	78
1986	86

Source: FHLBB data.

Table 6-9

FDIC and FSLIC Disposals[a] of Insolvent Banks and Thrifts, 1979–1988

	FDIC			FSLIC[b]		
	Number of Disposals	Disposals as a Percentage of All Insured Banks	Disposed Assets as a Percentage of All Bank Assets	Number of Disposals	Disposals as a Percentage of All Insured Thrifts	Disposed Assets as a Percentage of All Thrift Assets
1979	10	0.02%	0.01%	3	0.07%	0.10%
1980	10	0.07	0.01	11	0.28	0.23
1981	10	0.07	0.24	28	0.75	2.11
1982	42	0.29	0.53	63	1.91	2.52
1983	48	0.33	0.31	36	1.14	0.57
1984	79	0.54	0.13	22	0.70	0.52
1985	120	0.83	0.32	31	0.96	0.60
1986	138	0.97	0.24	46	1.43	1.07
1987	184	1.34	0.23	47	1.49	0.85
1988	200	1.47	1.04	205	6.95	7.45

[a] Instances in which an insolvent bank or thrift was disposed of, either through liquidation or through placement with an acquirer.

[b] For 1980 and after, only FSLIC-assisted acquisitions and liquidations are included; for 1979, supervisory cases are included as well.

Sources: Barth, Bartholomew, and Labich (1989); Barth, Feid, Riedel, and Tunis (1989); and FDIC data.

advantage of those opportunities. There were two additional events, however, that reinforced each other and that made an already bad situation considerably worse: changes in the tax laws and the rise and decline in the price of oil. Both contributed to a boom in commercial real estate, especially in the Southwest, and then its bust. Thrifts that financed this real estate were severely harmed.

The Economic Recovery Tax Act of 1981 shortened (for tax purposes) the depreciation periods that applied to real estate, thereby greatly increasing the relative profitability of real estate investments. Further, in the Southwest—especially Texas—the prospective profitability of real estate investments was enhanced by many investors' expectations of continuing increases in the price of oil. In the early 1980s the memories of the sharp rises in the price of oil in the fall of 1973 and again in the spring of 1979 were still fresh. Though domestic oil prices reached a peak of more than $34 per barrel in early 1981 and then began a gradual decline, many "experts" continued to predict that the price of oil could rise to $50, $80, or even $100 per barrel within the decade. Higher prices for oil would mean greater prosperity in the oilbelt, with multiple positive ramifications for real estate.

Fueled by favorable tax treatment and rosy expectations concerning oil prices, commercial real estate projects boomed—especially in the Southwest. Thrifts were major financers of and investors in these projects.

At the end of 1984 the U.S. Department of the Treasury issued a proposed revamping of the tax code. The proposal aimed at eliminating tax loopholes and other advantaged positions while lowering overall tax rates. The proposal—discussed and revised during the following two years—was finally passed by the Congress as the Tax Reform Act of 1986. Real estate was again a major area of change—this time in the opposite direction. Depreciation periods were lengthened, and the ability of real estate investors to shelter other income with their "passive" (depreciation-related) losses was restricted. This last provision was applied to existing real estate projects as well as to prospective ones. Such changes in the tax code—to the extent that they were unanticipated by the real estate markets before, say, 1985—could only mean a decrease in real estate values.

Movements in the price of oil reinforced this effect, at about the same time. After hitting their $34 per barrel peak in early 1981, oil prices drifted gradually downward during the next few years, settling to a $24 per barrel level in 1985. They then fell sharply in early 1986, dipping briefly below $10 per barrel during that summer. They subsequently rose modestly and fluctuated in a range around $15 per barrel. Thus, not only did the price of oil fail to rise to the heights that many had expected, but it fell to levels that were only half of those of the early 1980s. Again, this change could only mean a sharp fall in real estate values in the oilbelt.

Documenting the decline in real estate values in the Southwest is not an easy task, however. Though anecdotes abound, price series and indexes for real estate are rare, and those that do exist are frequently seriously flawed.[10] There is one real estate index, though, that does appear to satisfy the conditions for a reasonable indicator of real estate prices, at least for high-quality commercial properties. For more than ten years the Frank Russell Company (FRC) has collected quarterly data on the market values (as determined by appraisals) of the real estate under the management of a group of pension and trust funds.[11] The price data are available for four regions and four classifications of commercial property since 1977 and for eight subregions since 1982.[12]

Table 6-10 presents the FRC index for all of the properties in their sample across the entire United States, for their South region,[13] for their office building category across the entire United States, for their office building index for the South, and for all properties in their Southwest subregion.[14] The price index for all U.S. properties rose through 1985 and then declined modestly. The price index for the South reached a peak a year earlier and then fell by more than a fifth. The office building index for the entire United States peaked in 1985 and then declined by 10 percent. For office buildings in the South, however, the price index peaked in 1983 and then fell by over a third. Finally, the price index for all commercial properties in the Southwest sub-region—covering the states of Arkansas, Louisiana, Oklahoma, and Texas—peaked in 1983, slipped only slightly for two years, and then fell by almost a third.

The last two columns of Table 6-10 provide a price index series for crude oil and the price level of crude oil itself. Oil prices fell modestly after 1981 and then skidded sharply after 1985.

Since many real estate projects of the early 1980s were based on assumptions of a *continuing rise* in real estate values, even a leveling of prices would have had serious consequences for their viability. A severe price decline meant devastation for these projects—and for their investors and lenders. Thrifts were prominent among these two groups. Also, the real estate investments of the rapidly growing thrifts were probably of lower quality than those of the pension funds included in the FRC index, so it is likely that the property values of the former group fell earlier and by more than the declines that are indicated in Table 6-10.

Thus, the rapid growth of thrifts assets, much of it in new types of loans and investments, would have spelled serious trouble for the FSLIC in any event. The timing of the Tax Acts of 1981 and 1986 and of the decline in the price of oil, with their consequences on the prices of real estate in the South and Southwest, made a potentially bad situation substantially worse.

Table 6-10
Price Indexes for Commercial Real Estate[a] and for
Crude Oil, 1977–1989[b]

	All U.S.	All South[c]	Office Buildings U.S.	Office Buildings South	All Southwest[d]	Domestic Crude Oil Price Index	Domestic Crude Oil Price[e]
1977	70.4	73.5	59.9	n.a.	n.a.	31.6	$9.01
1978	75.2	26.5	66.2	n.a.	n.a.	32.6	9.28
1979	83.5	88.6	73.7	n.a.	n.a.	55.7	15.86
1980	91.1	89.8	86.7	n.a.	n.a.	85.5	24.35
1981	98.6	96.0	97.8	n.a.	n.a.	108.5	30.92
1982	100.0	100.0	100.0	100.0	100.0	100.0	28.49
1983	105.1	102.2	104.7	100.9	100.2	91.3	26.00
1984	110.8	104.9	110.0	100.3	100.1	89.4	25.47
1985	113.4	104.5	111.6	97.7	98.2	85.3	24.30
1986	112.6	98.9	108.7	88.9	90.5	40.0	11.36
1987	110.8	87.2	102.4	70.1	73.9	53.5	15.23
1988	111.0	83.8	98.8	64.2	69.8	38.5	10.97
1989	110.3	80.3	95.9	59.4	66.5	58.0	16.51

[a] Derived from the Frank Russell Company property index data.

[b] All data apply to fourth quarter prices for the years specified.

[c] Includes Alabama, Arkansas, Florida, Georgia, Louisana, Mississippi, Oklahoma, Tennessee, and Texas.

[d] Includes Arkansas, Louisiana, Oklahoma, and Texas.

[e] Per barrel.

Sources: Frank Russell Company and U.S. Department of Energy data.

Why the Delay in Reported Insolvencies?

As the previous sections have demonstrated, the true debacle occurred largely between 1983 and 1985; but the wave of losses, insolvencies, and closures began only in 1986. What explains this delay?

First, the standard (GAAP) accounting model is slow to recognize changes in the market value of assets. The accounting profession's focus on historical (book value) costs has created a mindset that has served well the purposes of thrifts (and other enterprises) that wish to delay the recognition of losses and give themselves more time to try to avoid the consequences of insolvency—possibly by "rolling the dice" a few more times.

Second, the GAAP framework permitted accounting practices that artificially bolstered short-run recorded income (and therefore assets and net worth) at the expense of long-run asset value. Thrifts could extend a loan to a developer that included funds to pay the interest on the loan for, say, the first two years; the developer would leave these future interest payments on deposit with the thrift. At the end of years 1 and 2, the thrift would reduce the developer's deposit by the size of the interest payment and record the interest "received" as income. In essence, the thrift was paying itself interest income out of the proceeds of the excessively large loan that the thrift had granted to the developer.[15] Another method was to charge the borrower large up-front fees, which could immediately be recorded as income, and then charge a lower (below market) interest rate on the loan (to compensate the borrower for the higher fees). The thrift could both record the fees as immediate income *and* enter the below-market loan as an asset at its face value.[16]

Through these and other "creative accounting" methods many thrifts were able to report higher short-run profits and net worth, at the expense of longer-run asset values. Thus, some of the apparently healthy profits shown by the thrift industry during 1983-1985 were accounting fabrications that would later unravel as growth ceased and property values tumbled.

Third, the Bank Board's establishment of RAP and its encouragement of the liberal creation of goodwill assets exacerbated the weaknesses of GAAP. Table 6-11 shows the numbers of thrifts that were insolvent on a RAP net worth measurement basis and the additional number (and their assets and net worth) that were insolvent on a tangible net worth measurement basis. As can be seen, after 1981 the RAP measurements portrayed as solvent hundreds of thrifts that were insolvent on the more stringent basis. It is important to emphasize that tangible net worth is not a perfect measure of true market value net worth, since it ignores some legitimate goodwill assets and also ignores subordinated debt liabilities (where the bondholders' claims are subordinated to the claims of the FSLIC) as part of the net worth that protects the insurer; however, it also relies on book value and ignores the market value of most tangible assets. The FSLIC, in disposing of insolvent thrifts, found that its costs usually exceeded the negative tangible net worth of a thrift, implying that the neglect of market values embodied in tangible net worth generally outweighed the neglect of legitimate assets and protections.

Thus, the effects of RAP in masking the negative tangible net worths (which themselves were underestimates of the eventual costs to the FSLIC) of hundreds of thrifts was a third source of the delay in the reported losses and insolvencies.

Fourth, as was shown in Table 6-10, real estate values began to decline precipitously only after 1985. Thus, though the loans and investments were made earlier, their declines in value occurred only in subsequent years; and, given the backward-looking framework of the accounting profession, these declines would be recognized on balance sheets even later.

Which Thrifts Failed?

The diversification of thrifts into new assets and new funding sources was not automatically suspect. *If done prudently, this diversification could have strengthened many thrifts.* Indeed, diversification was the intent of the legislative acts of the early 1980s.[17]

The rapid growth in amounts of these assets, however—especially by thrifts that were thinly capitalized—raises serious doubts about the motives of the thrift managements that were acquiring many of these assets.[18] A growing number of studies, using data gathered by the FSLIC as it disposed of insolvent thrifts in the late 1980s, indicate clearly that the nontraditional investments (and methods) of the rapidly growing thrifts of the 1983-1985 period were disproportionately responsible for the wave of insolvencies and their huge costs to the FSLIC.

One study examined the 205 insolvent thrifts that the FSLIC disposed of in 1988.[19] These thrifts (as compared with the remainder of the industry) were disproportionately composed of state-chartered institutions[20] and had grown significantly faster than the rest of the industry between 1983 and 1985. Their asset portfolios contained percentages of loans on land and of direct equity investments that were substantially above industry averages.[21] Initial (postmortem) investigations of their records indicated high percentages of suspected regulatory violations and fraud. In addition, all of these tendencies were yet more exacerbated for the 50 thrifts (among the 205) whose disposals were judged to be the most costly for the FSLIC.

A second study compared the asset percentages of (1) thrifts that either were disposed of between 1985 and 1988 or were RAP insolvent at the end of 1988 with (2) the remainder of the industry.[22] It found that the failed or failing thrifts had a disproportionately large percentage of their assets devoted to commercial mortgage loans, land loans, and direct equity investments.[23]

Table 6-11
Numbers, Assets, and Tangible Net Worth of Insolvent
Thrifts, 1981–1987

	RAP-Insolvent Thrifts[a]			RAP-Solvent but Tangible-Insolvent Thrifts[b]		
	Number	Assets (billions)	Tangible Net Worth (billions)	Number	Assets (billions)	Tangible Net Worth (billions)
1981	33	$3.3	$−0.1	77	$ 21.4	$−0.7
1982	71	12.8	−0.5	344	207.2	−12.3
1983	48	12.5	−1.6	467	272.1	−13.1
1984	71	14.9	−0.9	616	343.3	−15.9
1985	130	25.7	−2.7	566	326.7	−14.5
1986	255	68.2	−10.7	412	274.8	−9.1
1987	351	99.1	−21.0	321	255.4	−8.5

[a] Thrifts that were insolvent on the basis of regulatory accounting principles (RAP).
[b] Thrifts that were solvent on a RAP basis but insolvent on a tangible net worth basis.

Source: FHLBB data.

A third set of studies examined the determinants of the likelihood that a thrift would report insolvency between 1980 and 1988 and the determinants of the costs to the FSLIC of the insolvent thrifts that were disposed of during these same years.[24] These studies found that the likelihood of insolvency was positively related to a thrift's holdings of commercial mortgage loans, land loans, consumer loans, and direct equity investments in real estate.[25] The costs to the FSLIC were positively related to land loans, direct real estate ownership positions, and direct equity investments undertaken through a service corporation subsidiary.

Further evidence on the attributes of thrifts that failed in the late 1980s is provided in Table 6-12. This table compares the "nontraditional" asset holdings and net worth positions, *as of year-end 1985*, of two groups of thrifts: the group of thrifts (mentioned earlier in this chapter) that would subsequently fail,[26] and the remainder of the industry. This focus on 1985 allows us to examine asset percentages *before* significant writedowns were ordered by examiners in later years. As can be seen, the soon-to-fail group had appreciably higher percentages of all of the categories of nontraditional assets, except for consumer loans. They also had a higher goodwill percentage. At the same time, these thrifts were operating with lower levels of net worth, regardless of how it was measured. One further point is worth noting: The soon-to-fail group was disproportionately stock companies (by numbers and by assets) and disproportionately state-chartered (by assets).

The results of all these studies are generally quite consistent and point in a clear direction: The hundreds of rapidly growing thrifts of the 1982–1985 period did not use their new powers for prudent diversification. Rather, these thrifts plunged into new assets and investments in ways that increased their risks, not decreased them.

Fraud and Criminal Activity

The opportunities-capabilities-incentives nexus gave rise to expanded risk-taking in multiple dimensions. For some thrift managements, this expansion included deliberate violations of laws and regulations. Some violations involved inadequate underwriting and carelessness that contravened the fiduciary obligations of thrifts' managers and directors. Other violations involved loans to inappropriate individuals[27] (e.g., to managers, directors, owners, or other "insiders"; to their relatives; to businesses controlled by them; to their customers or suppliers; to their business associates) or in inappropriately large amounts.[28] Yet other violations involved fraudulent and excessive valuations of properties in order to justify larger loans to favored

Table 6-12
"Nontraditional" Assets and Net Worths of
"Soon-to-Fail" Thrifts and of the Remainder
of the Thrift Industry, 1985

	Soon-to-Fail Thrifts[a]	Remainder of the Thrift Industry
Number	669	2,577
Percent state-chartered	44.4%[b]	45.3%
Percent stock organization	60.7%[b]	27.0%
Assets (billions)	$286.8	$783.2
Percent state-chartered	62.6%[b]	27.8%
Percent stock organization	71.5%[b]	52.1%
Percentage of Assets:		
Commercial mortgage loans	13.4%	8.1%
Land loans	7.7	1.2
Commercial loans[c]	2.2	1.3
Consumer loans	4.4	4.2
Direct equity investments	5.0	1.7
Goodwill	2.8	2.1
Net worth as a % of assets:		
RAP measurement basis	2.6%	5.0%
GAAP measurement basis	1.4	4.0
Tangible measurement basis	-1.4	1.8

[a] Thrifts that either were liquidated or placed with acquirers during 1986–1989 or were predicted (as of early 1989) to require disposal.

[b] Includes MCP federal mutual thrifts that began 1985 as state-chartered stockholder-owned thrifts.

[c] Includes junk bonds.

Source: FHLBB data.

individuals and to justify larger asset values on the thrift's balance sheet (which, in turn, meant a larger net worth and the ability to grow or the ability to declare a dividend to the owners and thereby extract cash from the thrift).[29] Lavish spending on new corporate headquarters, executive salaries, client entertaining, company parties, and other excesses that were beyond reasonable business costs was still another type of violation.

Popular accounts of the thrifts' insolvencies—especially those in Texas, California, and Florida—have stressed these types of violations and have frequently left the impressions in readers' minds that these violations were the major reasons for all or most thrifts' failures and for losses generally.[30] A much-cited Congressional report did the same.[31]

There is no question that rules violations did occur in many thrifts. The anecdotes highlight egregious cases that clearly warrant substantial civil and/or criminal penalties, and there will surely be further instances of insider abuse or other repugnant behavior that future investigations will reveal. *The bulk of the insolvent thrifts' problems, however, did not stem from such fraudulent or criminal activities. These thrifts largely failed because of an amalgam of deliberately high-risk strategies, poor business judgments, foolish strategies, excessive optimism, and sloppy and careless underwriting, compounded by deteriorating real estate markets.* These thrifts had little incentive to behave otherwise, and the excessively lenient and ill-equipped regulatory environment tolerated these business practices for far too long.[32]

It is important that errant thrift executives be sued and prosecuted. Small amounts of moneys may be recoverable. The public's sense of justice—that wrongdoers are being punished—is important to satisfy. The deterrence of future wrongdoing through successful suits and prosecutions should be the goal of any legal system.

But any treatment of the S&L debacle that focuses largely or exclusively on the fraudulent and criminal activities is misguided and misleading. It perpetuates the incorrect notion, implicitly held by many, that virtually all the thrift insolvencies were caused by "crooks" whose ill-gotten gains are deposited in Swiss bank accounts (and if we could somehow find those bank accounts, we could recover all the moneys that are necessary to clean up the insolvencies). It also diverts attention from an understanding of how and why government policies went awry and distracts policymakers from the difficult but necessary policy reforms (discussed in Chapter 11) that are relevant for all of bank and thrift regulation.

Where Did the Money Go?

The subsequent insolvencies of the hundreds of "high-flyers" of the mid-1980s will require massive sums to be paid to honor the FSLIC's insurance guarantee to depositors. (More details are provided in Chapters 8 and 9.) A frequent question that is asked about these insolvencies is, where did the money go? An implicit assumption underlying this question is that "the money" must be "somewhere"—for example, in someone's safe deposit box in Switzerland.

This question confuses the losses that occur because of defalcations—employees' embezzling or absconding with bank funds—with those that are due to poor investments. In the former case, the act of the disappearance of the money is clear; and, if the employee has not yet spent the money, it is recoverable. (On the other hand, if the employee has spent it all on lavish travel and entertainment expenses, with no recoverable value, then "the money" has been dissipated and is gone; it may be traceable, but it is not recoverable.)

As argued earlier, defalcations (or their fraud-related equivalents) have not been a major part of the thrifts' losses. Some excessive dividends and salaries may have been paid. Some lavish lifestyles, accompanying the high incomes, may have been enjoyed. Some (relatively) small sums may have been squirreled away or may be protected by personal bankruptcy laws. Wherever possible these perpetrators should be sued and prosecuted, and all possible moneys should be recovered. But the overwhelming portion of the thrifts' losses did not occur in this way.

Where, then, did "the money" go when a thrift made a poor investment? A poor investment is one whose subsequent value is less than the value of the resources that were originally expended on it. Suppose a thrift bought 1,000 shares of the XYZ company at a price of $100 a share, hoping that the price would subsequently rise. Instead, the price falls to $60 a share. The thrift has lost $40,000 on its $100,000 investment. Where did "the money" go? It went to the sellers of the 1,000 shares. Unless there was something fraudulent about the transaction, the loss is traceable but not recoverable.

Suppose, instead, that the thrift invested in an office building that cost $10 million to construct. The thrift invested the $10 million (or made a loan to the developer for $10 million) because it expected the building to appreciate in value or because it expected that the future rents received from tenants would generate enough income to provide an acceptable return to someone who would buy the building at that appreciated price. But, contrary to expectations, the building declines in value to only $6 million—because tenants are fewer and rents are lower than had been anticipated. The thrift has lost $4 million. Where did "the money" go? In part, it went to the seller of the land; part went to the workers who constructed the building; part went to the suppliers of the concrete, the bricks, the windows, the plumbing fixtures, and so on. Again, in the absence of fraudulent transactions, the loss is not recoverable. In an important sense, the $6 million current value is the salvageable or recoverable value from the original $10 million investment; but the $4 million is lost.

This $4 million loss is a real social cost. Resources (land, labor, materials) that were valued at $10 million—that could have been devoted to producing alternative goods and services that would have had (and would continue to have) a value of $10 million—were devoted instead to this office building, which subsequently has a value of only $6 million. In retrospect, society would have received more value (satisfaction) from the resources if they had been devoted to the alternative uses. The losses of the S&L debacle are thus represented by the losses on thousands of ill-advised and inappropriate office buildings, shopping centers, condominiums, hotels, resorts, and other investments that were built with the thrifts' funds. The U.S. economy would have been better served if less resources had been devoted to these investments and more had been devoted to other uses.

The sad conclusion, then, is that "the money" has been dissipated and is not recoverable. It was largely invested in projects that were poor uses of the resources and have subsequently declined in value. It most certainly is not waiting in a safe deposit box in Switzerland.

Notes

1. This was true for operating losses as well as nonoperating losses. The writedowns of 1986 and after would generate nonoperating losses. And the reduction in or cessation of earnings from these written-down assets would generate operating losses.
2. The thrift industry had grown at average annual rates of 9.7 percent in the 1960s and 13.7 percent in the 1970s. But much of that growth was based on inflation-driven increases in house prices and mortgages.
3. The growth rates of these 637 thrifts include growth through merger, but merger was not the dominant form of growth. Further, even mergers put stresses on companies, as they have to absorb the new assets,

   operations, and personnel. Also, the growth of the remainder of the
   industry is overstated because interim de novo thrifts are included in the
   1985 asset figures, as are thrifts that switched into the FSLIC from state
   insurance funds in 1985 (e.g., from Ohio and Maryland).
4. Including mortgage-backed securities; MBSs have not been included in
   the "nontraditional" category because, for the most part, they represented
   straightforward claims on the underlying mortgages.
5. These percentages differ slightly from those shown in Table 2-7 because
   Barth, Bartholomew, and Labich (1989) use a slightly different measurement base from that used by Barth and Freund (1989).
6. Repurchase agreements, or "reverse repurchase agreements" as they
   are sometimes called, involved the thrift's "selling" a group of mortgages
   or mortgage-backed securities to a second party (typically, an investment
   banking house) and promising to "repurchase" at a future date, at a
   higher price. In essence, the thrift was borrowing from the investment
   banking firm (and repaying the loan with a larger sum that included
   interest) and pledging the mortgages as collateral.
7. The case of Lincoln Savings and Loan of Irvine, California, which
   achieved a great deal of notoriety in 1988 and 1989, was an instance of
   a new entrepreneur's buying a small stock thrift. For other stories
   involving the new entrepreneurs in the thrift industry, see Pizzo,
   Fricker, and Muolo (1989) and Adams (1990).
8. For discussions of the effects of economic deregulation in other industries, see Stoll (1979; 1981), Tinic and West (1980), Meyer et al. (1981),
   Keeler (1981), Bailey, Graham, and Kaplan (1985), Morrison and
   Winston (1986), Kaplan (1986), Moore (1986), and MacDonald (1989).
9. See Spellman (1982).
10. Most price series of real estate do not correct for changes in the types or
    quality of the properties for which the prices are being collected. For
    example, price series on the prices of single family homes usually do not
    hold constant the size, location, or attributes of the homes.
11. These properties may well be of higher quality, on average, than those
    in thrifts' loan portfolios.
12. These data are not based on prices revealed by transactions but, rather,
    on appraisals. It is likely that in a sagging market these appraisals lagged
    the prices that would have been revealed by transactions.
13. See Table 6-10 for the states in the region.
14. See Table 6-10 for the states in this subregion.
15. A thrift could achieve the same result by making a smaller initial loan
    and then loaning the developer the extra sums so that the developer
    could "pay" the interest when it was due.
16. Of course, the future income of the thrift would suffer because of the
    lower interest rate on the loan. But, if interest rates declined in the
    future, if more such fees could be earned in the future, or if the thrift's

other investments showed gains in the future, then these longer run losses would be washed away. A modification to the rules of GAAP in 1986 eliminated much of the abuse in this area of up-front fees.

17. See Garcia et al. (1983).
18. See Cole and McKenzie (1989).
19. See Barth, Bartholomew, and Labich (1989).
20. These state-chartered thrifts had an even more disproportionate representation in the costs that they imposed on the FSLIC.
21. There is a problem in measuring the percentages of asset types for thrifts that have failed: Since failure almost always involves the writedown of asset values, a "snapshot" of the thrift's balance sheet at the time of closure will reflect these writedowns and will not show fully the original investments in these faulty assets. An extreme example can illustrate this point: Suppose investments in asset X are subsequently found to be worthless and are written down to zero. A snapshot of the thrift's balance sheet at the time of closure would not reveal the previous investment in X and its role in the thrift's failure. The analysis presented in Table 6-12 avoids this problem by focusing on 1985, before significant writedowns began.
22. See Barth, Bartholomew, and Bradley (1989).
23. But consumer loans and commercial loans did not appear to be a problem.
24. See Cole (1990a; 1990b).
25. Direct equity investments undertaken through a service corporation subsidiary, however, were negatively related to the probability of insolvency; see also the articles cited in Note 15, Chapter 7.
26. The number of these soon-to-fail thrifts—669—that appears in Table 6-12 is larger than the 637 that are mentioned earlier in the chapter because the smaller number excludes thrifts that were not operating in both 1982 and 1985.
27. The regulations that cover these types of transactions are frequently referred to as the "transactions with affiliates" rules. They cover other types of transactions, in addition to loans, and include holding companies and their subsidiaries and affiliates as relevant parties.
28. There are limitations on the size of loan that can be made to a single borrower. The maximum loan to any single borrower was generally 100 percent of the thrift's net worth. The Financial Institutions Reform, Recovery, and Enforcement Act of 1989 lowered this maximum loan size to 15 percent of net worth.
29. Fraudulent appraisals and "land flips"—the repeated trading of properties between two or more thrifts, and successively higher (but false) values—were two such practices.
30. See Pizzo, Fricker, and Muolo (1989), Pilzer (1989), and Adams (1990).
31. See U.S. Congress (1988).

32. In many instances, the FHLBB regulatory officials would detect rules violations and send a letter demanding corrections. In the pre-1980 environment the letter would have been heeded. In the new environment, thrift managements would ignore, deflect, or delay the corrections, thereby gaining more time for the playing out of their risky strategies; see Pizzo, Fricker, and Muolo (1989, pp. 71 and 340) and Adams (1990, Chs. 11 and 12).

PART THREE

Regulatory Response and Cleanup

CHAPTER SEVEN

Tightening the Regulations and Recapitalizing the FSLIC, 1985–1987

Through all of 1983 and early 1984 the Federal Home Loan Bank Board was largely unaware of the extent of the destructive processes that the unbalanced economic deregulation of the previous three years had encouraged. Interest rates were declining from their previous highs. The thrift industry in aggregate was again showing profits after two years of severe losses.[1] Thrifts did seem to be growing their way out of their difficulties.

Suspicions, however, began to arise in early 1984. The Bank Board's closure of Empire Savings of Mesquite, Texas, in March 1984 began the reassessment.[2] Empire's problems had been created by shoddy loans and investments, not by the interest rate mismatch problems that had previously occupied the Bank Board's attention. Similar credit-quality problems began to surface in other thrifts. The wisdom of rapid growth began to be questioned.

After an ill-conceived effort to restrict brokered deposits, which was ultimately rebuffed by the courts, the Bank Board began to tighten the regulatory system. From late 1984 through the end of the 1980s the agency promulgated a broad range of regulations that included restrictions on rapid growth by thrifts with low net worth, higher net worth standards, reformed accounting, and limitations on direct equity ownership positions. Further, beginning in 1985, the Bank Board greatly expanded the numbers and quality of its field force of examiners and supervisors.

This regulatory tightening could not cure the problem loans and investments that had already been made by rapidly growing thrifts. Those loans were already embedded in the thrifts' asset portfolios; the losses that would eventually be recognized on them were now largely unavoidable.[3] The regulatory tightening was vital, however, to prevent yet more high-risk loans and investments from being made. In essence, though the barn door was being closed after far too many horses had escaped, the closing of the door was nevertheless crucial— so that even more horses would not escape. In hindsight, one could only wish that the door had been closed sooner, faster, and more effectively.

The actual and prospective insolvencies also meant the depletion of the FSLIC insurance fund. In early 1985 the Bank Board increased the deposit insurance premiums paid by thrifts by 150 percent. But the prospective drain on the fund was still beyond the immediate resources of the FSLIC. In late 1985 and early 1986, the Bank Board devised a legislative plan to "recapitalize" the insurance fund. Unfortunately, the Congress dithered for almost a year and a half, passing the necessary legislation only in August 1987. The delay itself, and the content of the final legislation (the Competitive Equality Banking Act of 1987, or CEBA), signaled clearly that the Congress did not yet believe that the insolvent thrifts posed a substantial problem.

Brokered Deposits: A False Start

The first efforts of the Bank Board to control thrifts' rapid growth, and the shoddy loans and investments that often accompanied it, involved restrictions on the insurance coverage for deposits gathered through brokers. With the end of Regulation Q (so that thrifts were not restricted in the interest rates that they could pay on deposits) and the increase in the insured amount to $100,000, a new group of financial services providers sprang into existence: deposit brokers. These brokers (many of them respectable Wall Street securities firms) would gather funds from individual investors and place them in large bundles (with the individual investors' names attached to $100,000 pieces within the bundle) in FSLIC-insured thrifts. Many rapidly growing thrifts used these brokered deposits as their primary source of funds to support their growth. As Table 6-5 showed, there was a direct correlation between thrift growth rates and their use of large denomination deposits, much of which were provided through brokers.

The Bank Board reasoned that the way to slow this growth was to restrict the flow of funds that was fueling it.[4] In January 1984 the Bank

Board proposed regulations that restricted insurance coverage to $100,000 for the aggregate of funds coming from any single source (e.g., a deposit broker), regardless of whether the funds were being deposited on behalf of multiple beneficiaries. The regulations were put into effect in April. The validity of the regulations was promptly challenged in court by a deposit broker. In June a Federal District Court judge issued an injunction to prevent enforcement of the regulations, and in January 1985 the U.S. Court of Appeals ruled that the regulation was indeed illegal and unenforceable.[5] This method of controlling growth had to be largely abandoned.[6]

The Bank Board's focus on brokered deposits was an unfortunate one—and not only because its efforts were eventually rebuffed by the courts. This effort was misguided for at least two reasons.[7] First, at a conceptual level, the brokered funds were not the direct reason for a poorly performing thrift's problems; rather, those problems stemmed from the poor loans and investments made by the thrift. Put another way, thrifts got into trouble because of where and how they *invested* their funds, not because of where and how they *gathered* those funds. The Bank Board was focusing on the wrong side of the thrifts' balance sheets.

Second, at a practical level, the regulation probably would not have damped the growth of many thrifts significantly. There were too many good alternatives to brokered deposits, and these alternatives would have been only slightly more costly for a thrift that was determined to pursue a high-growth strategy. Such thrifts could have advertised more extensively in the financial press; they could have relied (and many did) on internal "money desks" to contact investors directly in order to induce them to invest their funds in the thrift's CDs; or they could have relied on other sources of liabilities (e.g., "reverse repos"[8]). Clever investment bankers and other advisers to the thrift industry would likely have discovered yet other low-cost ways of channeling funds to thrifts that wanted them.

In sum, an attack on brokered funds was misguided on both conceptual and practical grounds. Thrifts that were intent on doing so could have found other low-cost ways to grow rapidly and to make their high-risk loans and investments. Brokered funds could be (and have been) used for benign purposes as well as the costly ones. It was the wrong place to focus regulatory attention.

Unfortunately, a bias against brokered funds (or "hot money") continues to pervade the bank and thrift regulatory arena.[9] This attitude either forces the institutions to use other, less efficient methods of gathering funds, or it condemns them to rely on funds that can only be raised through their local offices—-and it may thereby cause banks and thrifts to develop inefficient branch networks where brokered deposits are now efficient. It continues an outdated orientation toward localism when everything else in the banking environment is pointing toward regional, national, and even international markets.

One argument that is frequently used to defend restrictions on brokered funds can be paraphrased as follows: "If a bank or thrift pays high interest rates to attract brokered funds, it will be forced to invest those funds in high risk-endeavors, so as to earn sufficient income to cover those high interest costs." This argument misses the mark, however, because it reverses the logical causality. A bank or thrift is not likely to raise high-cost funds aggressively *and then* look for high-risk investments. Rather, it will have decided on a high-risk investment strategy and then look for the funding to finance that strategy.[10] And if it cannot use brokered funds, it will likely use some close (but slightly more costly) substitutes.

The proper regulatory focus should be on the asset quality and interest rate mismatch risks of a bank or thrift and on its net worth, not on where it gets its funds.

Strengthening the Regulatory Structure

After the false (and, ultimately, fruitless) start vis-à-vis brokered deposits, the Bank Board did focus on the correct parts of thrifts' balance sheets: their assets and their net worth. From late 1984 through late 1988 the Bank Board issued major sets of regulations that limited the ability of poorly capitalized thrifts to grow rapidly, that raised (albeit slowly) thrifts' net worth requirements, that limited the ability of state-chartered thrifts to take direct equity ownership positions, and that began a tightening of the accounting standards.[11] Also, through a clever exploitation of the public-private nature of the Federal Home Loan Bank System, the Bank Board was able to increase substantially the numbers, salaries, and quality of the field force of regulatory examiners and supervisors. Each of these areas will be addressed in the following discussion.

Limitations on growth. Since rapid growth in assets by poorly capitalized thrifts was seen as a major cause of the Bank Board's burgeoning problems, growth limitations were a logical place for regulatory emphasis. In early 1985 Bank Board regulations limited the annual growth of thrifts that were not meeting their net worth requirements; the ceiling growth rate was equal to the interest credited to depositors' accounts.[12] Thrifts that were meeting their net worth requirements could grow at rates of up to 25 percent per year without the permission of their supervisory agent (regulator) and at higher rates with permission. In 1986, with the change in the net worth standards (discussed next), thrifts that already exceeded the higher net worth goals were allowed to grow at rates faster than 25 percent per year, but were required to notify their supervisory agent of this rapid growth.

Net worth standards. In December 1984 the Bank Board issued regulations that specified that direct equity ownership positions needed to be accompanied by extra net worth levels. In the same month, the Bank Board began phasing out the five-year averaging requirement of the net worth standard, so that thrifts would have to meet the standard on an end-of-quarter "snapshot" basis.[13]

In September 1986 the Bank Board established higher net worth targets: 6 percent of liabilities, with a credit of up to 2 percent (i.e., as much as a 200-basis-point reduction) if a thrift satisfied a set of criteria that measured reductions in interest rate risk. This credit for interest rate risk reductions reinforced other efforts by the Bank Board, such as encouraging the originations of adjustable rate mortgages to reduce the thrift industry's exposure to movements in interest rates. The net worth targets were an eventual goal, with interim net worth standards rising at a gradual rate (from their 3 percent of liabilities starting level) that was geared to aggregate thrift industry profitability.[14] In an important sense, these net worth standards were sensitive to some of the risks in thrifts' portfolios, since they varied in relation to interest rate risk and to the level of direct equity ownership positions.[15]

In December 1988 the Bank Board proposed a more complete set of risk-related net worth requirements that called for net worth levels that were even higher and would be phased in faster. The proposed net worth requirements were modeled after those that had been adopted by the commercial bank regulators (the so-called Basel standards). The FIRREA legislation of 1989 superseded this proposal and required the issuance of a somewhat modified set of net worth standards in November 1989.

Direct equity ownership positions. The Garn–St Germain Act had limited the investments of federally chartered thrifts in direct equity ownership positions to 3 percent of their assets. There were no limits on this type of investments by state-chartered thrifts, however, and a number of states had given their state-chartered thrifts wide latitude in this area. During 1984 the Bank Board became convinced that these investments were a significant cause of losses for state-chartered thrifts and ultimately for the FSLIC.[16] In December 1984 the Bank Board approved regulations (which became effective in March 1985) that restricted the investment holdings of state-chartered thrifts and linked the percentage of their assets that could be devoted to these investments with these thrifts' net worths, as well as requiring extra net worth against these investments. In the spring of 1987 the Bank Board tightened the regulations by expanding the category of investments that were covered and linking the permitted percentages to a more stringent (tangible) net worth measure.

Reformed accounting. In the spring of 1987 the Bank Board passed regulations that began the phasing-out of RAP and the re-establishment of GAAP as the reporting standard for thrifts. The CEBA legislation of that year required a slight modification of that transition, which the Bank Board approved in December 1987.

Increased personnel. From the beginning of its chartering authority in 1934 the Bank Board had had an odd, bifurcated regulatory structure. The examiners—the personnel who performed the agency's audits of the thrifts—were directly on the agency's payroll and were civil service employees. But the supervisory personnel, who used the audit information for purposes of granting permissions or taking restrictive or disciplinary actions, were employees of the Federal Home Loan Banks and outside of the federal civil service system.[17] A "Chinese wall" in each FHLB separated these regulatory personnel from the bank's lending personnel,[18] and the president of each FHLB also had the title of principal supervisory agent (PSA). The bank president/PSA was responsible to the Bank Board for both the lending and supervisory sides of his bank, so ultimate government control over these governmental functions was preserved. But, with their employees outside the federal civil service system, the FHLBs had much greater flexibility with respect to staffing levels and salaries.

In late 1984, as the Bank Board recognized its need to increase its safety-and-soundness scrutiny, it asked the Office of Management and Budget (OMB) for authority for a substantial increase in the number of examiners. OMB countered with a far smaller increase that the Bank Board considered inadequate.[19] The Bank Board then devised a plan to evade OMB's control entirely: The examination staff was transferred to the FHLBs, thereby consolidating them with the supervisory staff (which made good organizational sense anyway) and removing them from federal government staffing and salary limita-tions. This strategy was carried out in July 1985. More examiners could now be hired, and they could be better paid and better trained. This latter point was equally important, since the federal civil service salary limitations had led to a deterioration in quality and high turnover among the examination staff.[20]

Simultaneously, the supervisory staff was expanded and upgraded. The combined effects of the expansion of examination and supervisory personnel can be seen in Table 7-1. Regulatory personnel and budgets expanded substantially after 1984 and by 1988 were far more than double their 1984 levels. Indeed, the Bank Board began to get a reputation as a "raider" of the personnel of other federal and state banking regulatory agencies.

The Consequences

By late 1986, with tighter safety-and-soundness regulations and an expanded (and expanding) regulatory field force, the Bank Board now had a firmer regulatory grip on the industry. Rapid growth was less common, and closer scrutiny of thrifts' investments and actions was more likely. With its expanded examination staff, the Bank Board increased the number of thrift examinations (audits) in 1986 by one-

Table 7-1
FHLBB Regulatory Resources, 1984–1988

	Examination and Supervision Staff[a]	Examination and Supervision Budget (millions)
1984	1,337	$67.0
1985	1,990	108.8
1986	2,986	168.5
1987	3,258	207.6
1988	3,411	226.2

[a] Full time equivalents, including non-professionals and support staff.

Sources: Barth and Bradley (1989); FHLBB data.

third, as compared with 1985.[21] A new focus on safety and soundness pervaded the regulatory system. Thrifts with low (or negative) net worths were put under much closer supervisory scrutiny and control, in order to prevent further risk-taking. The Bank Board was actively pursuing civil suits for breach of fiduciary obligations against errant managers and directors of thrifts, their accountants, their legal advisors, and their appraisers; and it was making criminal referrals to the Justice Department and urging indictments and prosecutions for criminal violations.

This new system was not perfect, however, and some thrifts (especially if the accounting conventions allowed them to show net worth levels that were above the required levels) managed to continue making high-risk loans and investments that would eventually be costly for the FSLIC. The higher net worth standards were being phased in quite slowly (the declining profitability of the industry would retard them yet further) and were still based on RAP; and the Bank Board had not embraced the fundamental reforms (discussed in Chapter 11) of market value accounting, risk-sensitive insurance premiums, and stronger powers to appoint a receiver. Still, the bulk of the abuse had been stopped—albeit far too late.

There are a number of indicators that support the view that by mid- or late 1986 the Bank Board had achieved much tighter (though not perfect) control over the thrift industry and that most of the abusive investment practices had slowed or stopped. First, the overall growth of the thrift industry was appreciably lower than it had been in the halcyon days of 1983 and 1984, and it was slowing. Growth of assets for the entire industry in 1986 was 8.7 percent; in 1987 it was 7.5 percent. These rates were less than half of those of 1983 and 1984.

Second, the industry was acquiring lesser amounts of new, "nontraditional" assets. As Table 7-2 indicates, the overall industry added only $14.4 billion in nontraditional assets to its books between year-end 1985 and year-end 1986; in the following year it added only half that amount, $7.1 billion.[22] By contrast, as Table 6-3 indicated, between year-end 1982 and year-end 1985 the industry had acquired $137 billion of these types of assets.

Third, of the group of thrifts discussed in Chapter 6 that were disposed of during 1986-1989 (or were predicted, as of early 1989 to require disposal), 609 were in operation at both year-end 1985 and 1987 (i.e., they had neither been liquidated nor placed with acquirers by the end of 1987). This group of 609 grew by only 10.0 percent over these two years, or less than 5 percent per year. Over half (55 percent) shrank in size. The thrifts that did grow were primarily those that were still reporting adequate net worths at that time.

Unfortunately, this tighter regulatory approach could not undo the costly mistakes of the previous few years. And, as the agency's examiners began increasingly to scrutinize and demand writedowns of questionable assets and as the real estate markets of the Southwest began to crumble badly, thrift losses mounted rapidly. With the mounting losses came increased numbers of thrift insolvencies. As Table 6-11 showed, even by the lenient standards of RAP net worth, the number of insolvent thrifts increased from 71 (with $14.9 billion in assets) at the end of 1984 to 130 (with $25.7 billion in assets) in 1985 and leaped to 225 (with $68.2 billion in assets) by the end of 1986.

The FSLIC could not possibly dispose of these growing numbers of insolvent thrifts in any short period of time because it lacked adequate funding and sufficient personnel.[23] Accordingly, the Bank Board needed a method for "warehousing" these "zombies" until

Table 7-2

Holdings of "Non-Traditional" Assets by FSLIC-Insured Thrifts, 1985–1987

	1985		1986		1987	
	Amount (billions)	Percentage of Total Assets	Amount (billions	Percentage of Total Assets	Amount (billions)	Percentage of Total Assets
Commercial mortgage loans	$98.4	9.2%	101.3	8.7%	$103.8	8.3%
Land loans	31.0	2.9	30.3	2.6	26.3	2.1
Commercial loans[a]	16.0	1.5	22.1	1.9	22.5	1.8
Consumer loans	43.9	4.1	48.9	4.2	55.0	4.4
Direct equity investments	26.8	2.5	27.9	2.4	30.0	2.4
Total	$216.1	20.2%	$230.5	19.8%	$237.6	19.0%

[a] Includes "junk bonds."

Source: Barth, Bartholomew, and Labich (1989).

adequate FSLIC resources for disposals were available. In many instances, tight supervisory controls were sufficient, as long as the insolvent thrift managements could be trusted not to try to evade the restrictions. But some managements could not be thus trusted, and an alternative mechanism was necessary.

In early 1985 the Bank Board devised the Management Consignment Program (MCP) for insolvent thrifts where the managements were unlikely to observe and obey the necessary supervisory controls. Since the thrift was insolvent, there were adequate grounds for declaring a "pass-through" receivership and removing the incumbent owners and managers. With those parties removed, the Bank Board would recharter the thrift as a federal mutual and install a caretaker group of managers, usually "consigned" from other thrifts in the state or region.

The first MCP thrift was created in April 1985, and the original vision of the program was that the thrift would remain in its MCP status for only ninety days; at that point the FSLIC would dispose of it. That first thrift—Beverly Hills Savings and Loan, of Beverly Hills, California—in fact remained as an MCP thrift for more than three-and-a-half years, with the FSLIC's finally finding an acquirer for it at the end of December 1988 (at an estimated cost of about $980 million). During the years 1985-1988 more than 100 insolvent thrifts were converted to MCP status and remained in this "holding tank" for greater or lesser periods. Many of this group were liquidated or placed with acquirers during 1987 and 1988, and at the end of 1988 only twenty-six MCP thrifts remained.[24]

Like the other regulatory control mechanisms available to the Bank Board at the time, the MCP managements could not miraculously convert bad assets into good ones and thereby regain solvency for the thrift. At best they could halt any new loans, foreclose on overdue loans, sue delinquent borrowers, liquidate some assets, let high cost deposits run off, and reduce operating expenses, while helping the FSLIC identify the value of the assets and thereby assist in the FSLIC's disposal efforts. (Asset value identification almost always meant writedowns and large nonoperating losses. Though these large losses were sometimes interpreted by the media as an indicator of poor performance by the MCP managers, in fact they were an indication that the managers were doing their job properly by identifying the embedded losses caused by the previous managements' poor decisions.) At worst, because the MCP managements were paid flat fees that did not reward performance (which, itself, was usually not adequately specified), they might serve as pure caretakers and allow existing loans and investments to deteriorate further in value.

In almost all instances, though, the MCP thrifts found that they had to pay higher interest rates on their deposits than would have been true if they had been solvent.[25] Despite FSLIC insurance and the FSLIC's consistent record of paying off insured depositors in liquidated thrifts immediately after closure, some depositors were leery of leaving their funds in an insolvent thrift and required a higher interest rate to compensate them for their perceived risk. (The experience of depositors in some state-chartered thrifts in Ohio and Maryland in the spring of 1985, where the state-sponsored deposit insurance funds failed and depositors saw their funds frozen and eventually lost a fraction of their deposits, surely encouraged this wariness.) These higher interest rates added to the operating losses to the MCP thrifts and also had ripple effects on healthy thrifts in adjoining areas, forcing them to raise their interest rates on deposits and thereby raising their costs.[26]

The continued operation of the MCP thrifts (and of dozens—and then hundreds—of other insolvent thrifts) was a salient indication of the FSLIC's inadequate resources to dispose of all of the problem thrifts in its growing caseload. More resources for the FSLIC were vital.

Recapitalizing the FSLIC

The FSLIC's sole sources of income were the flat-rate insurance premiums levied on thrifts' deposits, plus interest earned on the FSLIC's accumulated funds. Since 1950 the annual premium had been 1/12 of 1 percent of a thrift's deposits, or 8.33 cents per $100 in deposits. In 1984 the FSLIC's premium income was $597 million, and its interest receipts were another $728 million, for a total income of $1.325 billion. Its net reserves[27] at the end of 1984 were $5.6 billion.

By early 1985 it was clear that the growing numbers of insolvent thrifts were quickly overwhelming the FSLIC's monetary and personnel capabilities for disposing of them. Quick closure of just the 71 RAP-insolvent thrifts at the end of 1984 would have required about $15 billion to be paid out to insured depositors and other secured creditors, which was far in excess of the FSLIC's $5.6 billion in reserves. Though liquidations of the assets of these institutions would recover some of the moneys, the liquidations could not occur instantly. The sale of the assets would extend over a few years and would surely yield substantially less than their $14.9 billion nominal value. Even if the assets could somehow be sold quickly, the net cost to the insurance fund—and hence the reduction in its net reserves—would be substantial. In addition, the FSLIC's annual premium income plus any interest on the remaining shrunken pool of FSLIC's assets of the fund would not be nearly large enough to handle the even larger number of insolvencies that were likely to occur.

The alternative disposal mechanism—placing thrifts with acquirers—would conserve on the FSLIC's initial outlay. Since acquirers would take on the liability to the insolvent thrifts' depositors, the FSLIC's outlay would "only" have to be amounts that were roughly equal to the difference between the amount of these deposit liabilities and the smaller market value of the insolvent thrift's assets (including any "franchise" value of the thrift as a going concern). But this net cost would be about the same (or a little less) than the net cost of liquidations, and the reduction in the FSLIC's reserves would be substantial. Large-scale disposals of the insolvents, by either method, would require substantially more resources.

The Bank Board's first move was to increase the deposit insurance premiums paid by thrifts. The National Housing Act permitted the Bank Board to levy a "special assessment" of 1/32 of 1 percent of deposits per three-month period. For an entire year, this came to 1/8 of 1 percent, or a 150 percent increase in the premium level. The Bank Board first began levying this special assessment in March 1985, and for all of 1985 the regular premiums brought $704 million into the FSLIC fund, and the special assessments yielded an additional $1,011 million.

The rising tide of insolvencies, however, was clearly swamping these revenue-raising efforts. Despite nine liquidations and twenty-two placements of insolvent thrifts with acquirers during 1985, at an estimated aggregate cost to the FSLIC of $1.0 billion, the number of RAP insolvent thrifts had grown to 130 by the end of 1985, with $25.7 billion in assets. The Bank Board had felt the necessity to convert twenty-five insolvent thrifts to MCP status. More resources were desperately needed.[28]

The Bank Board and the Reagan administration in late 1985 and early 1986 crafted a plan for raising more resources.[29] They operated under a major constraint: Though the FSLIC was an agency of the federal government, it had always supported itself financially through insurance premiums that were levied on thrifts' deposits (as was true for the FDIC and banks); moneys from the U.S. Treasury (i.e., taxpayers) had never been involved. All of the parties wanted a mechanism that maintained the principle of the FSLIC's being a self-supporting fund of the thrift industry and that avoided a Treasury commitment and an increase in the reported budget deficit, especially at a time of very large federal budget deficits.

The plan that was developed, which was dubbed the "recapital-ization" of the FSLIC, entailed the following: The FSLIC would borrow up to $15 billion in thirty-year bonds through a new entity: the Financing Corporation (FICO). The interest on these bonds would be paid from future deposit insurance premium revenues (including the special assessment). The repayment of the principal would be guaranteed by the immediate commitment of up to $3 billion from the net worths of the twelve Federal Home Loan Banks (whose shareholders were the thrift industry[30]). The compounded earnings of these committed funds would easily yield enough to achieve $15 billion in thirty years.[31] Because the Treasury was not involved in either the payment of interest or repayment of principal, these bonds would not be considered "full faith and credit" instruments and would carry a higher interest rate (which was a cost to the FSLIC);[32] however, they thereby avoided being counted as part of the federal deficit and the national debt.

In essence, the FSLIC was being authorized to borrow $15 billion against a stream of future premiums. It was telescoping those future premiums into the present, in order to get the resources to use immediately to dispose of its caseload of insolvent thrifts. The only real, net addition of resources that had not previously been available to the FSLIC was the commitment of the funds from the FHLBs, which freed up resources for the FSLIC that it otherwise would have had to commit to the repayment of the bonds rather than to the disposing of the insolvents.

This program was presented to the Congress in April 1986. The Congress felt no sense of urgency and proceeded to consider it at a leisurely pace. The lack of urgency stemmed from at least four sources. First, the inability of the FSLIC to liquidate or place insolvents with acquirers was not seen by the Congress as a crisis. Depositors were not, except in a few instances,[33] running on these thrifts to withdraw their funds. These insolvent thrifts continued in operation, and their continued losses seemed remote and unimpor-tant. Second, neither the Bank Board nor the Reagan administration presented the problem to the Congress or to the public as one of crisis. This was done, at least partly, to avoid creating headlines (e.g., "Thrift

Deposit Insurance Fund is Broke") that might scare depositors and lead to runs on insolvent and solvent thrifts alike. The memories of massive runs in 1984 on Continental Illinois Bank and on American Savings of Stockton, California, and of runs on the state-chartered thrifts in Ohio and Maryland in 1985, were still fresh. Also, the full depth of the problems of the troubled thrifts—still masked by an accounting system that was slow to recognize market values and by the Southwest real estate markets that were only beginning to crumble— were not yet apparent to the agency or to the Treasury leadership.

Third, many in Congress did not understand the plan, nor the nature of the FSLIC's commitments to depositors and its difficulties from thrift insolvencies, and simply saw the plan as more government borrowing and spending, which they wished to avoid. Fourth, some saw the troubles of the insolvent thrifts as just another cyclical phenomenon that time and waiting would cure. Most of the thrifts that had been squeezed by the high interest rates of the early 1980s had regained profitability when interest rates declined after 1982. The same might happen for the insolvent "class of '86." This last approach both misunderstood the poor quality of many of the loans and investments of the insolvents and underestimated the losses and risks to the FSLIC from the continued operation of the insolvents.

As the FSLIC recapitalization bill languished in the Congress during the summer of 1986, a new argument for delay arose. The operators of Texas thrifts began complaining to their Congressmen, and especially to the Majority Leader of the House Jim Wright (who would become Speaker of the House six months later), that the Bank Board regulators were being too harsh and were inhibiting their business operations. More money for the FSLIC would make it easier for the Bank Board to close more of them down. Wright responded by placing a hold on further progress of the bill.[34]

The hold was eventually lifted, but valuable time was lost. Though both the House and Senate did pass versions of the bill that would have authorized the $15 billion in borrowing for the FSLIC, their bills had differing provisions that required reconciliation and agreement in a conference committee. The two houses could not agree, and the Congress adjourned in October without passing the legislation.

By the end of 1986 it was clear that the troubled sector of the thrift industry was getting worse. Though the solvent portion of the industry remained profitable, losses were mounting for the insolvents and soon-to-be insolvents. During 1986 the Bank Board disposed of another forty-six insolvent thrifts—ten through liquidations and thirty-six through placements with acquirers—at a total estimated cost of $3.1 billion. Nevertheless, the number of insolvents (measured by RAP) mushroomed to 255, with $68.2 billion in assets, by the end of 1986. These RAP-insolvent thrifts lost $6.3 billion during 1986, of which $2.9 billion was operating losses and $3.3 billion was in nonoperating (writedown) losses.

In January 1987 the FSLIC recapitalization bill was reintroduced in both houses, but again no sense of urgency accompanied it. Further, a strong sentiment had developed in the Congress (encouraged by thrift industry lobbying) that the Bank Board's recent regulatory actions, including thrift disposals, had been erratic and that the Bank Board should be kept on a short Congressional leash. This sentiment manifested itself in a number of ways. First, as before, there was no urgency to pass the bill. Second, instead of the $15 billion requested by the Bank Board, the Senate version of the bill authorized only $7.5 billion of borrowing for the FSLIC, with a maximum of $3.75 billion to be borrowed in any twelve-month period; the House version, in response to a strong lobbying effort by the main thrift trade association (the U.S. League of Savings Institutions), authorized only $5 billion in total borrowing. Third, the House version required the Bank Board to "forbear" and allow thrifts that had very low net worth ratios—as little as 0.5 percent of liabilities—to continue to operate, provided that the thrifts met other "good faith" tests.[35] In a major sense, this forbearance provision was redundant, since insolvency was almost always the grounds that the Bank Board used for appointing a receiver—the action that would trigger the removal of the previous owners and managers—and the Bank Board already had a caseload of more than 250 insolvent thrifts that it would have to dispose of. Still, it was indicative of the Congress's mood, which favored leniency.

During the spring of 1987 the reported losses and insolvencies continued to mount; and the General Accounting Office (the Congress's auditing agency for the FSLIC) declared that the FSLIC, as of year-end 1986, was insolvent because the "contingent liabilities" of the insolvent (but not yet disposed of) thrifts exceeded the FSLIC's assets. Also, as the FSLIC continued to dispose of insolvent thrifts and honor other obligations in early 1987, its liquid assets fell below $1 billion. With its interest income gone and its insurance premium income limited to approximately $500 million per quarter, the FSLIC could now only dispose of a few, small insolvents and would not be able to handle anything larger. Still, the Congress dithered, and the bill languished.

At this point Bank Board and Treasury officials were in a bind. Excessively loud entreaties to the Congress that the FSLIC was broke might cause depositor nervousness and runs, while still not springing loose the bill—the worst of all possible worlds. Yet, in the absence of such entreaties, the Congress did not seem to feel any sense of urgency.

Finally in late July, after the Reagan administration threatened to veto a bill that contained inadequate borrowing authority, a compromise was reached: The bill authorized $10.825 billion in borrowing,[36] but the $3.75 billion limit on borrowing in any twelve-month period remained, as did the forbearance provisions. The bill did instruct the Bank Board to re-establish GAAP as the reporting system for thrifts by 1993, which would appear to have been a call for regulatory tightening. But the language of the specific provisions, plus the accompanying explanations of the conference committee, make clear that the Congress (responding to the complaints of the Texas thrifts) believed that the Bank Board had been acting more harshly with respect to asset writedowns than was warranted under GAAP and that an insistence on GAAP represented more lenient treatment for thrifts.

In sum, in the summer of 1987 the sense of the Congress—strongly influenced by the lobbying of the thrift industry—was for forbearance and leniency and for tight limits on the Bank Board's ability to dispose of the insolvents. The bill embodying these provisions, the Competitive Equality Banking Act of 1987, was signed by President Reagan on August 10.

Would Earlier Congressional Action Have Made a Difference?

In the "finger pointing" that accompanied the discussion of the FIRREA legislation of 1989 and the much larger estimated costs of dealing with the insolvent thrifts, some critics on and off Capitol Hill have claimed that earlier action by the Congress in passing the CEBA—say, passage a year earlier, in the summer of 1986—would have reduced those costs considerably. The spirit of this criticism clearly is correctly aimed. The Congress did take far too long to pass the CEBA, and passage of the 1986 version would have meant the full $15 billion in borrowing authority, no twelve-month restrictions on the amounts borrowed, and no forbearance and leniency provisions. But the cost savings would have been relatively modest.

With the full $15 billion available to it, the FSLIC could have disposed of more thrifts, and especially could have liquidated more thrifts, earlier and thus would have halted the hemorrhaging of their operating losses sooner.[37] The real savings in costs to the FSLIC, however, would have been only the difference between the operating losses that the insolvents actually incurred as they continued to operate and the borrowing costs that the FSLIC would have incurred on the FICO bonds. The much larger cost to the FSLIC were the losses in asset values—the writedowns—of the insolvent thrifts. These assets were already on the books of the thrifts, and thus the losses—either already recognized or still to come—were largely unavoidable for the FSLIC, regardless of whether the FSLIC disposed of those thrifts in 1986, 1987, or later.[38]

An example will illustrate this last point. Suppose, to take an extreme possibility, that the entire $15 billion authorized by the 1986 version of CEBA could have been borrowed by the end of 1986. Suppose, further, that the $15 billion would have been adequate to have disposed rapidly of two-thirds of the 255 thrifts that were insolvent on a RAP measurement basis at the end of 1986.[39] Those 255 thrifts had operating losses of $2.9 billion in 1986, so disposals of two-thirds of them would have reduced annual operating losses by approximately $2 billion.[40] But the FSLIC's annual interest costs on the $15 billion of borrowing would have been about $1.5 billion, so the net savings from earlier action would have been "merely" $0.5 billion.

If earlier action by the FSLIC would have meant forestalling further risk-taking (and eventually costly) actions by these thrifts, the cost savings could have been higher. But by 1986 these insolvents were under supervisory control or had been converted to MCP status. Though a few insolvent thrift managements might have slipped a few additional costly actions past their regulatory supervisors, *the major source of the FSLIC's costs were the bad loans and investments that these insolvent thrifts—and the additional hundreds of thrifts that would slide into insolvency in the following three years—had already made in earlier years; by 1986 it was too late for the FSLIC to cut its losses by much.*[41]

Thus, as is now known, even $15 billion in borrowing authority for the FSLIC in the summer of 1986 was wholly inadequate for the depth of the problem. Rapid passage of the CEBA would have helped a little bit and would have avoided the misguided forbearance and leniency provisions of the 1987 version. But it could not have saved large sums.

Notes

1. As noted in Chapter 6, however, some of the apparent profits were accounting devices that allowed thrifts to bolster short-run profits at the expense of longer-run asset values.
2. See Pizzo, Fricker, and Muolo (1989, Ch. 7) and Adams (1990, Ch. 11).
3. The losses could have been avoided only if the FSLIC had liquidated the thrifts and quickly sold the assets to private investors at their inflated prices, so that the private investors would have borne the eventual losses. This was not a likely scenario.
4. Support for this position came from the leadership at the FDIC.
5. See *FAIC Securities Inc.* v. *U.S.,* 595 F. Supp. 73 (1984), 753 F. 2d. 166 (1985), 762 F. 2d. 352 (1985).
6. A modest limitation on the use of brokered deposits—no more than 5 percent of deposits by thrifts that were below their minimum net worth requirement—remained.
7. See also Benston (1984; 1985)
8. This was a means by which a bank or thrift could pledge its assets in return for low cost loans from securities dealers and other funds providers.
9. This attitude found its way into the FIRREA, which forbids any bank or thrift that is not meeting its net worth requirements from accepting brokered deposits. In defense of this provision, it does focus attention on the thinly capitalized thrifts, where the risks are highest; and it does give the FDIC flexibility to override the prohibition. Still, it continues to focus too much attention on the wrong side of the balance sheet.
10. See Spellman (1982, Ch. 3).
11. Many other regulations were issued as well. Partial compilations are found in Kane (1989) and Barth and Bradley (1989).
12. Depositors usually leave in their accounts the interest that accrues on their deposits. This source of deposit expansion does not imply any aggressive growth behavior.
13. In November 1983 the net worth requirements for de novo thrifts had been raised to 7 percent of liabilities for their first year.
14. At least part of the reason for gearing the increase in required net worth to profit levels was to give mutual thrifts, who could not raise capital by selling shares of stock (unless they were prepared to convert from their mutual status), time to increase their net worth through retained earnings. The shrinking of mutual (and other) thrifts by letting

deposits run off and selling assets, in order to increase the net worth to liabilities ratio, was not considered to be a serious option.

15. There has been a continuing controversy as to whether these direct equity investments were risky and contributed to the likelihood of a thrift's becoming insolvent and to higher costs to the FSLIC in the event of an insolvency; see Benston (1986), Barth, Brumbaugh, Sauerhaft, and Wang (1989), Benston (1989b), Benston and Koehn (1989), and Brown, McKenzie, and Cole (1990); see also the evidence discussed in Chapter 6.

16. The rhetoric surrounding the hostility to these "direct investments" was sometimes less focused on safety-and-soundness considerations and more on the claim that it was inappropriate for insured deposits to be invested in specific types of commercial enterprises (e.g., fast food franchises). Why it was appropriate for banks and thrifts to use insured deposits to make commercial *loans* (funded by insured deposits) to these enterprises but not to make direct equity investments (aside from safety-and-soundness considerations) was never explained.

17. Their salaries were part of the costs of the FHLBs and thus reduced the FHLBs' profits and dividends that otherwise would have been remitted to their thrift owners.

18. Also, the directors of each FHLB—two-thirds elected by the thrift owners, one-third appointed by the Bank Board—were expected to observe the "Chinese wall" and devote all of their attentions only to the lending operations of their bank.

19. See Pizzo, Fricker, and Muolo (1989, Ch. 22) and Adams (1990, Chs. 11, 13).

20. See FHLBB (1984).

21. See FHLBB (1987).

22. One caveat to these figures should be mentioned. In 1986, the thrift industry's nonoperating losses—capital losses and writedowns—were $4.8 billion; in 1987 they were $8.9 billion. Unfortunately, the asset categories to which these losses applied are not available. Since the figures in the text are net of losses, the gross additions of "nontraditional" assets could have been somewhat higher. However, even if all of the nonoperating losses involved these asset categories (which was unlikely), the gross additions of these assets to the thrifts' balance sheets would still have been far less than during the 1983-1985 period.

23. Additional personnel were required, in the event of thrifts' liquidations, to supervise the payouts to depositors or the transfers of the deposits to other thrifts and to manage and dispose of the assets acquired by the FSLIC. In the event of placements of insolvent thrifts with acquirers, the personnel were needed to negotiate the terms of the FSLIC's payments to the acquirers and to monitor the subsequent performance of the acquirers under the terms of the agreements.

24. Two more MCP thrifts were disposed of in early 1989. Also, in 1988 the FSLIC "stabilized" eight thrifts in Texas and ten thrifts in Oklahoma that had not been disposed of by the end of 1988.

25. Studies by Hirschhorn (1989), Cook and Spellman (1989), and others have consistently shown that the interest rates paid by banks and thrifts (other things being held constant) vary inversely with the institution's net worth.

26. A likely transmission mechanism to explain this ripple effect would be as follows: To attract deposits from knowledgeable, risk-averse depositors who were fearful of an insolvent thrift's demise and of the FSLIC's ability to honor its insurance guarantee, the insolvent thrift would have to pay higher interest rates. Since the thrift could not distinguish or discriminate among its depositors, it would have to pay these higher rates to all its depositors, including those who were less knowledgeable about or insensitive to the thrift's financial condition or to the FSLIC's condition. The competition for this latter group's deposits would then cause the healthy thrifts to raise their deposit rates. For the evidence indicating that thrifts generally had to pay higher interest rates as the financial position of the FSLIC weakened, see Hirschhorn (1989) and Cook and Spellman (1990).

27. Its assets less its liabilities or, in essence, its net worth.

28. Also during 1985, in anticipation of the need to manage and dispose of large amounts of assets—loans in various conditions of arrears, as well as real estate holdings—that it would come to possess through liquidations of thrifts, the Bank Board established the Federal Asset Disposition Association (FADA). The FADA was a private entity that could pay private sector salaries and avoid federal government personnel limitations; however, it would still be answerable to the Bank Board. Ideally, the FADA would have achieved the best of both worlds: It would carry out governmental functions, but be able to attract the specialized skills in asset management and disposal that required private sector salaries. Unfortunately, little thought was given to the FADA's specific mission and role or to its reporting and control relationships with the liquidation division of the FSLIC. Also, the early leadership of the FADA was insensitive to the political nuances of Washington. The FADA was controversial from its beginning and gained a great deal of Congressional hostility. Though the FADA eventually did become an effective manager and disposer of assets, its reputation and the continuing hostility by the Congress proved to be too much. The FIRREA legislation instructed the FDIC to liquidate the FADA, which was done in early 1990.

29. The twelve FHLB presidents/PSAs played an active role in the development of this plan.

30. In the FHLBs of Boston, New York, Pittsburgh, and Seattle were some shareholder members that were FDIC-insured MSBs.

31. The specific mechanism was to buy a thirty-year Treasury zero coupon bond, which was used to "defease" or guaranty the repayment of the principal.

32. The higher interest rate—the differential between thirty-year FICO bonds and thirty-year Treasury bonds—was initially approximately 0.80 percent (80 basis points). It later ranged as high as 1.10 percent (110 basis points) and as low as 0.50 percent (50 basis points).

33. There had been a major run on the largest thrift in the United States, American Savings and Loan of Stockton, California, in the fall of 1984.

34. See Jackson (1989); Phelan (1989); Pizzo, Fricker, and Muolo (1989, Ch. 17); and Adams (1990, Ch. 12).

35. The Bank Board, in hopes that a concession to the Congressmen who favored forbearance would speed passage of the recapitalization legislation, had issued a policy statement in February 1987 endorsing forbearance for thrifts with thin (but positive) net worth. The policy statement closely tracked one that had been issued for national banks a year earlier by the Comptroller of the Currency; and, in light of the large number of insolvent thrifts that were already in the FSLIC's caseload, the policy statement was likely to have little net effect on the FSLIC's immediate actions vis-à-vis these not-yet-insolvent thrifts. Still, in retrospect, it was the wrong signal to send.

36. The extra $825 million, above the round number of $10 billion, was authorized so that the Bank Board could refund to thrifts some extra contributions (the "secondary reserve") that they had made to the FSLIC fund in the 1960s and early 1970s.

37. Also, to the extent that a few of these insolvent thrifts were able to evade supervisory scrutiny and were able to continue to acquire risky (and, ultimately, costly) assets, earlier action would have reduced the FSLIC's costs commensurately; however, by 1986 the scrutiny and controls over the RAP insolvents were reasonably tight.

38. See Note 3. If the FSLIC had found acquirers for the insolvent thrifts, the acquirers would have had only limited ability (their net worth investments in the thrifts) to absorb the losses on the assets that were declining in value, and then the thrifts would have been back in the insolvent column.

39. The 255 insolvent thrifts had $68.2 billion in assets and negative tangible net worths of $10.7 billion. The estimated cost to the FSLIC of disposing of them was substantially above the latter number because writedowns of their assets were still occurring. Though $15 billion would not have been adequate to handle the immediate cash needs for the liquidation of even two-thirds of this group ($50 billion in immediate

cash would have been necessary), the $15 billion probably would have been adequate to place two-thirds of them with acquirers (and perhaps could have included a few small liquidations).

40. A proper triage system would have focused on disposing of the heaviest "bleeders" first, so the reduction in operating losses might have been slightly greater. Further, since the financial condition of insolvent thrifts (including their operating losses) inevitably deteriorated with time, the savings from rapid action might have been slightly greater for this reason as well. Also, more rapid action would have lowered the deposit costs for other thrifts; see Note 26.

41. Some thrifts that were reporting net worth levels in 1986 that were above the regulatory minimums clearly did pursue risky strategies that eventually caused their insolvencies and extra costs to the FSLIC in 1988 or 1989; but more funds for the FSLIC in 1986 would not have prevented these outcomes because the FSLIC would have used the funds to dispose of its insolvent caseload at that time. Only a better (market value) accounting system, higher net worth standards, and tighter regulatory scrutiny could have avoided these costs.

CHAPTER EIGHT

Beginning the Cleanup, 1988

With the passage of the Competitive Equality Banking Act (CEBA) in August 1987, the Congress had grudgingly acknowledged that some cleanup activity vis-à-vis the insolvent thrifts was necessary. But the Congress clearly did not expect much activity, since it had restricted the FSLIC's formal borrowing through thirty-year bonds to a total of $10.825 billion, with a further limit of only $3.75 billion per year. As a number of critics noted, this latter amount was barely enough to dispose of only the insolvent thrifts that one would find "along the highway from the airport to downtown Dallas."

At the same time, the fraction of the thrift industry that was insolvent was growing, and its losses were mounting. The industry had split into two economic groups: a solvent majority that was making moderate profits, and a deeply insolvent minority, whose losses were swamping the former's profits. For all of 1987 the aggregate thrift industry's operations yielded record losses.

In this environment, the Federal Home Loan Bank Board began its efforts to dispose of as many insolvent thrifts as it could. The agency hired additional personnel to deal with the liquidations and with placements with acquirers. It began to borrow the authorized funds through the Financing Corporation (established by the CEBA). A special plan for dealing with the insolvent thrifts of Texas was developed. Though the CEBA had limited the FSLIC to borrowing $3.75 billion per twelve-month period through the FICO, the FSLIC was able to develop other, innovative financing mechanisms that allowed it to stretch its resources considerably farther. During 1988 the FSLIC disposed of 205 thrifts, at an estimated present discounted cost to the FSLIC of about $30 billion.[1]

These disposal transactions quickly became highly controversial and widely misunderstood. They mostly involved placements of thrifts with acquirers. The sums that the FSLIC was required to commit—because of the deep insolvencies of these thrifts—were large. The reason why the FSLIC would have to pay an acquirer to take an insolvent thrift was not well understood, nor were the financing and payment mechanisms that the FSLIC used. Some of the acquirers—notably, William Simon, Robert Bass, and Ronald Perelman—were controversial financial figures in their own right. The FSLIC took advantage of the tax reductions that acquirers might receive, in order to allow the FSLIC to stretch its resources further. The media repeatedly used the term "bailout" in referring to these transactions—and even to liquidations—and the term has stuck, though it carries misleading connotations concerning the nature of these transactions.

Preparing for 1988

In passing the CEBA in August 1987, the Congress had hoped that the problems of the insolvent thrifts were temporary or cyclical and would soon pass. As 1987 came to an end, however, it was clear that the troubles of the insolvent portion of the thrift industry were growing larger, not smaller. As Table 6-10 indicated, at the end of 1987 the number of RAP insolvent thrifts had grown to 351. These thrifts had assets of $99.1 billion, and they had negative tangible net worths of $21.0 billion. The aggregate net income reported for the overall thrift industry in 1987 was a record loss of $7.8 billion.

This loss was both an overstatement and an understatement of the problems of the industry and of its insurer. On the one hand, the loss was masking an underlying bifurcation of the thrift industry: Two-thirds of the industry was still profitable; only one-third was running losses. This segmentation pattern was different from that of the industry's previous period of losses, 1981 and 1982, when virtually the entire industry was running losses. As can be seen in Table 8-1, however, the large operating and nonoperating losses of the un-profitable thrifts in 1987 were swamping the profits of the profitable thrifts. Media headlines through 1987 and 1988 focused on the overall losses of the industry and conveyed the impression that all of the firms in the industry were drowning in red ink, rather than a minority.

On the other hand, what mattered from the perspective of the FSLIC was the gross losses of the insolvent or soon-to-be insolvent thrifts, not the net losses that resulted from the netting of the healthy thrifts' profits against the insolvents' losses. Those gross losses added to the FSLIC's liability, and the profits of the healthy thrifts did not offset that liability for the FSLIC.[2] Of the 1,106 unprofitable firms that incurred the $14.4 billion in gross losses, 672 were insolvent on a tangible net worth basis; these 672 accounted for more than five-sixths ($12.1 billion) of the gross losses and had aggregate negative tangible net worths of $29.5 billion. The FSLIC's liability was clearly mounting.

Despite its meager resources, the FSLIC had continued to dispose of insolvent thrifts during the first half of 1987. The passage of the CEBA provided some bolstering of the FSLIC's resources and, after a few months' lag, the agency again began disposing of insolvents. For all of 1987 the FSLIC liquidated seventeen insolvent thrifts and placed another thirty with acquirers, at an estimated cost of $3.7 billion. As Table 8-2 indicates, in the four years of 1984 through 1987 the FSLIC

Table 8-1
Profits and Losses of FSLIC-Insured Thrifts, 1985–1988

	1985	1986	1987	1988
All Thrifts:				
Net Income (billions)	$3.7	0.1	−7.8	−13.0
Profitable Thrifts:				
Number	2,558	2,350	2,041	2,024
Assets (billions)	$908.1	901.0	828.4	906.2
Net operating income (billions)[a]	$5.8	8.9	8.7	7.5
Net nonoperating income (billions)[a]	$3.8	3.5	1.0	0.4
Net income (billions)[b]	$7.3	9.0	6.6	5.6
Unprofitable Thrifts:				
Number	688	870	1,106	925
Assets (billions)	$161.9	262.9	422.5	444.5
Net operating income (billions)[a]	$−2.2	−4.3	−5.9	−6.8
Net nonoperating income (billions)[a]	$−1.6	−4.8	−8.9	−12.2
Net income (billions)[b]	$−3.6	−8.9	−14.4	−18.6

[a] Before taxes.
[c] After taxes.

Source: FHLBB data.

Table 8-2

The FSLIC's Liquidations and Placements with Acquirers of Insolvent Thrifts, 1980–1988

	Liquidations			Placements with Acquirers			Total		
	Number	Assets (billions)	Estimated Cost[a] (billions)	Number	Assets (billions)	Estimated Cost[a] (billions)	Number	Assets (billions)	Estimated Cost[a] (billions)
1980	0	$0.0	$0.0	11	$1.5	$0.2	11	$1.5	$0.2
1981	1	0.1	0.03	27	13.8	0.7	28	13.9	0.7
1982	1	0.04	0.003	62	17.6	0.8	63	17.7	0.8
1983	5	0.3	0.1	31	4.4	0.2	36	4.6	0.3
1984	9	1.5	0.6	13	3.6	0.2	22	5.1	0.7
1985	9	2.1	0.6	22	4.2	0.4	31	6.4	1.0
1986	10	0.6	0.3	36	11.9	2.8	46	12.5	3.1
1987	17	3.0	2.3	30	7.6	1.4	47	10.7	3.7
1988	26	3.0	2.8	179	97.7	27.1[b]	205	100.7	29.9[b]

[a] Costs to the FSLIC on a present discounted value basis; neglects costs to the U.S. Treasury in the form of reduced tax collections.

[b] Reduced tax collections by the U.S. Treasury were estimated to be an additional $5.5 billion, on a present discounted value basis.

Sources: Barth, Bartholomew, and Bradley (1989); FHLBB data.

had disposed of 146 insolvent thrifts, with aggregate estimated costs of $8.5 billion. These costs were more than four times the dollar amounts of those that had been incurred in the first four years of the decade.

Nevertheless, the number and magnitudes of insolvent thrifts had continued to mount. More, and more rapid, disposals were necessary.

Further, the insolvent thrifts of Texas had emerged as a special problem. Table 8-3 presents the comparison of the Texas thrifts with the rest of the country. As of the end of 1987 the insolvent thrifts (measured on a RAP basis) of Texas accounted for almost one-third (31.1 percent) of all RAP-insolvent thrifts in the country and well over one-third (39.6 percent) of the assets in all RAP-insolvent thrifts. These percentages were far in excess of the overall representation of Texas thrifts in the national thrift industry. Also, the unprofitable thrifts in Texas accounted for more than half of the losses of all unprofitable thrifts throughout the country. Of the twenty thrifts with the largest losses in 1987, fourteen were in Texas.

This information concerning the insolvencies and losses of the Texas thrifts was not lost on their depositors. Though depositors in thrifts generally had become skittish about the health of their thrifts and of the FSLIC, this effect was felt acutely in Texas. A "Texas premium"—a difference of 0.50 percent (50 basis points) or higher that Texas thrifts had to pay on deposits, as compared with what thrifts elsewhere were paying, just to retain those deposits[3]—had developed.

Table 8-3
Comparisons of FSLIC-Insured Thrifts in Texas and in
the Remainder of the U.S., 1987

	All Thrifts		RAP-Insolvent Thrifts	
	Number	Assets (billions)	Number	Assets (billions)
Texas	279	$99.3	109	$39.2
Rest of U.S.	2,868	1,149.6	242	59.9
All U.S.	3,147	1,248.9	351	99.1

Source: FHLBB data.

This Texas premium added to the operating costs of insolvent and solvent thrifts alike. With respect to the former group, these extra costs added directly to the FSLIC's eventual costs of disposing of these insolvents; as to the still-solvent thrifts, these extra operating costs reduced their profits or added to their losses and drove them closer to insolvency and to being an eventual burden to the FSLIC. With $85 billion in deposits in Texas thrifts at the end of 1987, even a 0.50 percent differential meant a difference of more than $400 million per year in net income.

Clearly, a special effort was required for Texas.

The Strategy and Its Outcome

The strategy for disposals of insolvent thrifts by the FSLIC, as it evolved in late 1987 and early 1988, was as follows: The cash that was raised from insurance premium revenues, from the sale of assets (which had come into the FSLIC's possession from earlier liquidations of insolvent thrifts), and from the flotation of thirty-year bonds through the FICO would be used largely for liquidations. Liquidations were cash-intensive, since either a direct payout to depositors or a transfer of the deposits to another thrift required immediate cash. Only in subsequent years, as the assets of the liquidated thrifts were sold, would some of the cash be replenished.

Placements of insolvent thrifts with acquirers consumed much less of the FSLIC's immediate cash resources, for three reasons. First, even if the acquirer demanded cash as the asset that was necessary to fill the insolvent thrift's negative net worth "hole"—the difference between the shrunken market value of the insolvent thrift's assets and its deposit liabilities—that amount would be appreciably less than the initial cash distribution to depositors in a payout. Second, by preserving the "going concern value" of a thrift through a placement with an acquirer—in essence, by preserving an intangible asset that would be lost if the thrift were liquidated—the FSLIC lowered its costs of disposal and hence the cash needed. Third, and most important for conserving the FSLIC's cash, acquirers were willing to accept notes issued by the FSLIC and to accept other promises of future payment by the FSLIC as "hole-filling" assets in lieu of cash. These substitute assets were not possible in payouts to depositors or in most transfers of deposits to other thrifts.

These notes and other promises of future payments by the FSLIC were an important financing mechanism for the FSLIC. They allowed the FSLIC to stretch its resources considerably farther, and thus to dispose of considerably more insolvent thrifts. In essence, like borrowings through the FICO, these financing mechanisms allowed the FSLIC to borrow against future revenue streams and telescope its future resources into 1988. It could thereby dispose of many more thrifts than if it were limited to its $2 billion per year in premiums, the $3.75 billion per year permitted in FICO borrowings, and any asset sale proceeds from earlier liquidations.

The "Southwest Plan" that was developed for the disposal of the insolvent Texas thrifts had, at its heart, these financing mechanisms. In addition, the vision of the plan was to achieve a shrinkage and consolidation of the Texas thrift industry. Though Texas was 6.8 percent of the U.S. population and accounted for 7.2 percent of U.S. GNP, Texas thrifts accounted for 8.9 percent of all thrifts by number and 9.1 percent of all thrift deposits. Acquirers would be expected to acquire groups of adjacent or complementary insolvent thrifts, thereby allowing the acquirer to close overlapping branches and consolidate operations. Deposits would shrink as high-cost deposits were allowed to run off and were replaced with lower cost advances (loans) from the Federal Home Loan Bank of Dallas, and the FSLIC notes and other financial promises to the acquirer would provide the collateral for these FHLB advances. These actions, reinforced by the public's perception that the FSLIC was prepared to deal with the Texas insolvents and by a separate program administered by the FHLB of Dallas to replace high-cost deposits in insolvent Texas thrifts with lower cost deposits gathered from other thrifts,[4] were expected to reduce or erase the Texas premium, lowering the costs for all Texas thrifts.

The last piece of the FSLIC's strategy was to make use of the tax reductions that acquirers would enjoy. The various tax acts passed by the Congress in the 1980s had restricted the ability of acquirers of loss-ridden companies generally to reduce their (the acquirers') taxes by taking advantage of the previous embedded losses of those target companies; however, the Congress had preserved this privilege for acquirers of insolvent thrifts, as well as exempting them from paying taxes on the interest on the notes and promises issued by the FSLIC in connection with the placement of an insolvent thrift. As the losses in insolvent thrifts grew larger, these tax aspects became more important; also, these exemptions were scheduled to expire at the end of 1988. In essence, these potential tax reductions for acquirers were an extra asset that could be attached to an insolvent thrift, thereby reducing the necessary "hole-filling" payments from the FSLIC. These exemptions were thus a way of substituting money from the Treasury (forgone tax collections) for FSLIC resources and allowing the FSLIC to stretch its resources farther.

The execution of this strategy resulted in a large volume of FSLIC disposals of insolvent thrifts. As Table 8-2 indicates, during 1988 the FSLIC liquidated 26 thrifts and placed another 179 with acquirers (in 86 separate acquisition transactions). The impending expiration (at the end of 1988) of the full[5] ability of acquirers to use the tax reduction benefits for deals concluded after December 31 caused a flurry of transactions in December: 75 insolvent thrifts (out of 179 for the entire year) were placed with acquirers in 35 transactions concluded in December.

The 205 thrifts that were disposed of during 1988 had assets of $100.7 billion; their disposals were estimated to cost the FSLIC $29.9 billion.[6] In addition to these costs, the FSLIC estimated that the Treasury would lose another $5.5 billion in foregone tax collections.[7] Of the 205 insolvents that were disposed of, 81 were Texas insolvents.[8] The costs of disposing of these 81 Texas thrifts were estimated to account for more than 60 percent of the FSLIC's $29.9 billion disposal costs for the year.

These transactions were highly controversial and generated considerable criticism. Their sums were large, and the necessity for committing such large sums was not well understood; the terms of the transactions were complicated; and some of the acquirers were controversial individuals. Even the terminology surrounding these transactions impeded a better understanding of their basic logic and necessity.

Disposing of an Insolvent Thrift: The Basic Features

As an aid to understanding the 1988 transactions, it is useful to return to the balance sheet diagrams of Chapter 3 and, more specifically, to the balance sheet of an insolvent thrift. Table 8-4 reproduces the balance sheet of the insolvent thrift shown in Table 3-4. The assumption that was made for expositional purposes in chapter 3—that the values shown on the balance sheet represent market values—will be continued here, unless otherwise indicated.

The essence of a thrift's insolvency is that the value of its assets is inadequate to cover the value of its liabilities. In Table 8-4, that shortfall is $32. Since the deposit insurer has promised the depositors that they will receive their full $92, the deposit insurer will necessarily incur a loss that is approximately equal to that shortfall of $32.

This last point is true because there are basically only two ways for the insurer to dispose of an insolvent thrift and thereby satisfy its insurance obligations to depositors: liquidating the thrift or placing it with an acquirer. In a liquidation the deposit insurer pays out the $92 to the depositors and simultaneously liquidates the assets.[9] As a variant on this direct depositor-payout form of liquidation, the insurer could transfer the deposits to another institution; however, to offset these liabilities, the insurer would have to give the acquiring depository $92 in cash or equivalent assets. In either variant, the deposit insurer pays out $92 on the deposits but receives only $60 from liquidating the assets.[10] The insurer's net cost is $32.

A liquidation requires the immediate payout of cash, whereas the actual liquidation of the assets takes time—typically, five years or more are required before all of the assets of a liquidated thrift or bank are sold or otherwise realized (e.g., through the eventual repayment of borrowers' loans). This differential in the timing of the payout and eventual receipt of cash by the deposit insurer means that the immediate cash ($92) that is required for the liquidation payout is considerably larger than the net cost ($32).[11] Also, the delayed receipt of the proceeds of the sale of the assets must imply that the $60 in asset value is the arithmetic result of using an interest rate to discount a somewhat larger stream of asset sales in the future years; equivalently, if the sale of the assets in future years yielded only $60 at those times, their market value at the time of closure would be less than $60 (and

Table 8-4
The Balance Sheet of an Insolvent Thrift, as of
December 31, 199Z

Assets	Liabilities
$60 (loans)	$92 (deposits, insured)

	–$32 (net worth)

the net cost to the insurer—on a present discounted value basis—would be greater than $32).

The second method for disposing of the insolvent thrift is to place it with an acquirer—say, another thrift, a bank, or a company or individual that would like to enter the thrift business. The acquirer, though, is receiving $92 in liabilities (obligations to depositors) and only $60 in assets. The acquirer will insist on approximately[12] $32 in additional assets (cash or other) from the deposit insurer in order to have the value of the assets it receives be equal (approximately) to the value of the liabilities it receives. Again, the deposit insurer's net cost is $32.[13]

In this second scenario, the deposit insurer's need for immediate cash is equal to its net cost and is considerably less than in the liquidation scenario. If the acquirer is willing to accept promises of future payments (which have a discounted present value of $32) from the deposit insurer, then the latter's immediate cash needs are reduced yet further.

If immediate cash availability were not a constraint for the deposit insurer, and if the asset values realized were identical for the liquidation or acquirer scenarios, the deposit insurer should be indifferent between the two. But, in either event, unless the deposit insurer intends to "stiff" the depositors, it cannot avoid the $32 loss.

At least two considerations, however, would tend to lower the deposit insurer's costs for the acquisition scenario and thus favor that outcome.[14] First, placement with an acquirer would preserve any "going concern" or "franchise" value of the thrift: the value of its customer loyalty, its brand-name recognition, its business relationships with its depositors and loan customers. This franchise value might be minimal for thrifts that had gathered only "jumbo" (around $100,000) deposits through brokers or money desks and hence did not have significant "core deposits" (i.e., a group of loyal depositors) and had made mostly (or entirely) shoddy or fraudulent loans.[15] This franchise value could be significant, however, for thrifts that had substantial core deposits and that had a mixture of good and bad assets. A liquidation involving a payout to depositors would dissipate all of this value. Even a liquidation involving a transfer of the deposits to another thrift or a bank would preserve only the value of the depositor relationships. In essence, any going-concern value of a thrift was an "off-balance-sheet" asset that would reduce the insurer's cost of filling the thrift's negative net worth "hole" and could be preserved fully only through placement of the thrift with an acquirer.

Second, achieving the highest value for the insolvent thrift's assets might best be accomplished by leaving them in private hands (i.e., the acquirer's) where the motives of direct financial gain from enhanced value would apply most forcefully. The liquidation alternative would involve federal government employees as the liquidating receiver and thus as the asset managers and disposers. Federal salary ceilings made difficult the employment of highly skilled real estate professionals, who could command higher salaries in the private sector. Additionally, with federal employees as managers and disposers, the motives of direct financial gain from enhanced asset values were largely absent.[16] It appeared that the private sector could do a more effective job and thus reduce the deposit insurer's costs. In essence, the recoverable (market) value of the assets of an insolvent thrift were likely to be higher (and thus the deposit insurer's net costs would be lower) in the hands of an acquirer than in the hands of a (government) liquidating receiver.

The Structure of the FSLIC's 1988 Transactions

Since placement of an insolvent thrift with an acquirer was generally the lower cost route, the FSLIC almost always first tried to "market" an insolvent thrift. The field of potential acquirers was quite large. Unlike commercial banks—which could not be owned by firms involved in nonfinancial commerce or industry, in insurance, or in investment banking (securities)[17]—almost any individual, group, or company could own a thrift provided they met the requisite standards of good character, financial capability, and managerial competence. In fact, thrift owners have encompassed companies as diverse as Ford Motor Company, Fuqua Industries, Weyerhaueser, ITT, Gulf & Western, Household International, and Sears, Roebuck. In addition, the Garn–St Germain Act of 1982 made it possible for commercial banks to acquire insolvent thrifts, even if the latter were based in a different state from the former (and hence the acquisition would otherwise have contravened the existing state prohibitions on inter-state branching by banks and thrifts).

The FSLIC developed lists of prospective acquirers, made presentations, held seminars, and generally tried to promote the acquisitions of these insolvents. Lists of the available insolvents were circulated. In the case of Texas, potential acquirers were urged to bid for groups or clumpings of insolvents, so as to achieve cost savings from consolidations.

Where there was no acquirer interest in a thrift, that thrift would be relegated to the likely liquidation list.[18] Where one or more potential acquirers for a thrift showed interest, bidders' conferences were held, and negotiations ensued to achieve the lowest cost to the FSLIC. Each acquirer's bid (in terms of the size, form, and timing of the payments from the FSLIC that the acquirer would consider acceptable) had to be evaluated by the FSLIC on the basis of its estimated present discounted cost. That cost had to be lower than the FSLIC's estimated present discounted cost of liquidating the thrift[19]; otherwise, the thrift would be relegated to the liquidation list. The interest rate that the FSLIC used for these discounted cost calculations was the FSLIC's cost of borrowing on its thirty-year FICO bonds—approximately 10 percent.

If the market value of the insolvent thrift's assets were readily ascertainable (say, the $60 of Table 8-4), a simple payment to fill the negative net worth "hole" (the $32 of Table 8-4) by the FSLIC—in the form of cash or a ten-year note from the FSLIC,[20] or a mixture of the two—would be sufficient. For many insolvent thrifts, however, especially those in Texas, determining the market values of many of their assets would have been extremely time-consuming, costly, and plagued with unknowns.[21] An example would be a loan in arrears on a partially completed suburban Texas shopping center or office complex that was also tied up in litigation. An acquirer would understandably be leery of any valuation that might quickly be placed on that asset; and, if forced to do so, the acquirer would surely place a low value on it. In the terms of Table 8-4, the insolvent thrift's assets might have historical book values of $60, but their market values were likely to be lower (but difficult to ascertain quickly), and a risk-averse acquirer might value them at only $25, necessitating a much larger "hole-filling" payment by the FSLIC.

To deal with this last problem, the FSLIC developed contracts that provided "capital loss coverage" on these "covered assets": The assets would be transferred to the acquirer at nominal book value ($60 in the example), with a promise by the FSLIC to make up the differences between their eventual sale (or other disposal) values and their book values. The contracts sometimes included some loss sharing arrangements with the acquirer and always included gain sharing, so that the acquirer would have an incentive to enhance the value of the assets; however, the FSLIC would retain the lion's share of any recovery in asset value. Additionally, in the interim (until sale or disposal), the FSLIC provided "yield maintenance" payments on the assets (many of which yielded little or no current income) so that the acquirer could cover its deposit interest and other operating costs until the time when the asset was sold.[22]

In essence, these capital loss coverage and yield maintenance arrangements allowed the FSLIC to substitute future payments for immediate payments and to defer placing a value on these hard-to-value assets. Through these arrangements the FSLIC retained most or all of the downside risk regarding the financial outcome of these assets; however, it also retained most of the upside gain that might accrue. The alternatives looked even less palatable: To transfer all of the risk to the acquirer would have involved lower transfer prices for these assets and higher "hole-filling" payments by the FSLIC; these low transfer prices might well be revealed subsequently to have been too favorable to the acquirer.[23] Or the FSLIC could have retained the assets and instead provided the acquirer with cash or notes (equal in value to the book value of the assets—$60 in Table 8-4), but then the FSLIC would still bear all of the downside risks and would face all of the costs and difficulties of managing and disposing of the assets itself.

In short, the dilemma of where and how to manage these "difficult" assets had no easy solution.[24] Leaving them in the hands of acquirers was chosen as the best alternative.[25]

Two other aspects of these transactions are worth noting. First, the FSLIC insisted on the inclusion of "call" provisions on most of the notes that it issued and on most of the covered assets that had capital loss coverage and yield maintenance provisions. If, in the future, the FSLIC (or its successor) became more flush with resources and decided that paying off the notes early and/or acquiring the assets and managing and disposing of them directly was worthwhile, it could do so.

Second, in all of the Texas transactions (and a few outside Texas), the FSLIC received warrants that it could convert into a percentage—usually 20 percent[26]—of the stock of the new thrift.[27] Thus, if the new thrift prospered, the FSLIC's warrants would increase in value and reduce the FSLIC's net cost of the transaction.[28]

The tax code treated the interest on the note from the FSLIC[29] and the yield maintenance payments as nonreportable income, which would reduce the taxes of the acquirer. In addition, the earlier annual losses of the insolvent thrift could be carried forward and used to offset the future profits of the successor thrift and thus again reduce tax payments. These tax benefits were recognized as extra assets. In many transactions the FSLIC negotiated terms that included the direct assignment of all or most of these tax savings to the FSLIC (i.e., the FSLIC's payments to the acquirer were reduced by the amounts that the acquirer saved in lower tax payments); in others, the FSLIC received a lower percentage of the tax benefits but was able to negotiate more favorable terms on other parts of the transaction.

With the inclusion of the FSLIC's hole-filling payment the value of the assets balanced the value of the liabilities; the delivered thrift was no longer insolvent. The acquirer would be expected to infuse enough cash or other resources into the thrift so that these additional assets provided it with an adequate net worth. The new thrift would be subject to tight scrutiny by the Bank Board's enhanced (see Chapter 7) field force of examiners and supervisors. In some instances waivers of some restrictive regulations were granted for limited periods of time. In the new environment of close regulatory scrutiny, however, these waivers were considered to entail acceptably small risks to the FSLIC. Also, in most transactions the FSLIC negotiated a "pre-nuptial" clause that allowed it to take control of the successor thrift at a specified *positive* level of net worth, thus giving the FSLIC an extra layer of protection.

Were the FSLIC's Transactions "Bailouts"?

The media continually referred to the FSLIC's disposals of insolvent thrifts during 1988 (and earlier) as "bailouts" or as "rescues."[30] The FIRREA legislation of 1989 was commonly referred to as "the S&L bailout bill," and the term has stuck. Bailout carries the connotation of a financial windfall or unexpected gain—usually, relief from financial difficulties—for a group that is undeserving. The constant impression conveyed is that the moneys are yet another example of unwarranted government spending and subsidies that only benefit a group of special interests: "the S&Ls."

The term is unfortunate, because it carries the wrong implications and overtones as to the nature of these transactions. It is not a correct characterization of these disposals of insolvent thrifts. *The moneys expended were going, and will go, to satisfy the U.S. government's obligation to insured depositors, either directly in liquidation payouts to depositors or indirectly by finding acquirers to assume the obligations to depositors.*

Because "bailout" has become a pervasive term, however, further analysis is warranted. If, contrary to the previous statement, these transactions were bailouts, the beneficiaries would have to be people; a reference just to "the insolvent thrifts" is not an adequate description, because the concept of bailing out bricks and mortar makes little sense. It is *people* who would have to be the recipients of the windfall or unexpected gain of a bailout. But a review of the principal parties connected to these transactions confirm that the term is inappropriate and a mischaracterization of these disposal actions.

Owners, directors, and managers of the insolvent thrifts. Under a legal framework of limited liability, the owners of any corporation normally cannot be asked to contribute more than their ownership stake (net worth); after their ownership stake has been exhausted, they normally bear no further financial responsibility to the liability holders for covering the insolvency.[31] The same principles apply to the owners, directors, and managers of insolvent thrifts. It is true, as Chapters 5 and 6 documented, that inappropriate public policies in the mid-1980s allowed this group generally to retain their positions and pursue their costly business strategies for far too long. Many enjoyed high incomes and the accompanying life styles. Some may even have squirreled away some comparatively small sums that may not be recoverable in civil suits (e.g., because of personal bankruptcy protections). But virtually all of the accumulated shortfall in the insolvent thrifts is not recoverable in any form, because the loans and investments made by these thrifts are no longer worth their original amounts.

As their now-insolvent thrifts were being disposed of, through liquidations or placements with acquirers in 1985 and after, *none* of the FSLIC's moneys were being received by this group. In all instances the ownership interests of the previous owners were erased; their financial stakes in these enterprises were irretrievably lost. Further, the acquirers rarely retained the former senior managers.[32] And, where civil or criminal violations of these former owners, directors, or managers have been detected, they have been sued or indicted. The disposals of the insolvents in the late 1980s did not bail out this group.

Lower-level employees of the insolvent thrifts. Some of these employees retained their positions in the successor thrifts, though many lost their jobs through consolidations or liquidations. Tellers or lower-level loan officers were not responsible for these thrifts' insolvencies. They were not being bailed out.

Insured depositors. They had a promise from or "contract" with the

FSLIC, covering them with deposit insurance (up to $100,000). The moneys are being used, directly or indirectly, to satisfy that promise. If an insurance company pays off the beneficiaries on a promised insurance arrangement, it is difficult to describe that as an "unexpected gain or windfall."

As Chapters 5 and 6 demonstrated, this contract was woefully mismanaged by the Congress, the Reagan administration, the Federal Home Loan Bank Board, and many state governments. Still, the promise was made. Making good on that promise cannot be considered a bailout.

Uninsured depositors. The amount of uninsured deposits—amounts above $100,000—in insolvent thrifts was typically only 1-2 percent of all deposits. In the event of liquidations, any holder of a "jumbo" deposit received only the $100,000; amounts above $100,000 became just another set of claims on the assets in the liquidating receivership.[33] If an insolvent thrift was placed with an acquirer, the uninsured amounts above $100,000 were sometimes transferred to the acquirer and sometimes subjected to the insurance limitations; the choice was based on other considerations that determined which route was least costly to the FSLIC.[34] In the former case, one might legitimately claim that these few uninsured depositors were being bailed out;[35] but the sums were, both relatively and in aggregate, small.[36]

Owners of solvent thrifts. The Bank Board increased the insurance premiums paid by all FSLIC-insured thrifts in 1985, and the FIRREA calls for another increase in 1991. It is the healthy thrifts that pay these premiums. The CEBA eliminated a small part of their ownership stake in the Federal Home Loan Bank System; the FIRREA taxed away a larger share. The FIRREA also reduced their flexibility of operations. These actions are the antithesis of a bailout.

Only if one believed that these healthy thrifts had an implicit obligation to make good on all shortfalls that the FSLIC might experience, and that even the higher premiums that they have been and will be paying do not adequately satisfy that obligation, could one then claim that the owners of solvent thrifts are being bailed out. Since the cleaning up of all of the insolvent thrifts—the costs of honoring all of the insurance obligations to depositors in these thrifts—greatly exceeds the net worth (the owners' stakes) of the solvent thrifts, the effort to satisfy any such obligation would eliminate all current owners of solvent thrifts. Again, it is difficult to claim that this group is being bailed out.

Acquirers. If acquirers struck particularly favorable deals with the

FSLIC, such that they received payments that were larger than the minimum they would have accepted to acquire the thrift, they may thereby have received a set of "windfalls." (This point will be addressed in the next section.) But any such excess payments to acquirers were surely not synonymous with bailing out the insolvent thrifts.

A second terminological problem is associated with these payments to acquirers, and this one is of the FSLIC's own making. The technical term that the FSLIC used for these payments was *assistance payments*.[37] The contract with an acquirer was an *assistance agreement*. The transaction was an *assisted acquisition*. Sometimes the transaction was described as the "sale" of the insolvent thrift, with "FSLIC assistance."

The FSLIC's payments, in reality, were just the necessary filling in of the shortfall between the inadequate value of an insolvent thrift's assets and the value of its liabilities—a cost that the FSLIC would have to bear in any event. These payments were "assistance" only in the sense that an acquirer would not have accepted these thrifts without these payments.

To the uninitiated, however, the term *assistance* looked (and continues to look) suspiciously close to *subsidy*. The inclusion of tax breaks for acquirers as part of the transactions exacerbated these suspicions, and so the term *subsidies* was sometimes used in news stories to describe the FSLIC's payments to acquirers. Though this usage also created misimpressions and incorrect overtones, at least it recognized that the moneys were going to the acquirers and not to former owners. Some media stories, however, would discuss the bailout of the thrifts and the payment of subsidies to acquirers in adjacent paragraphs, without recognizing the contradiction between the two terms.

In sum, the FSLIC's actions and payments in disposing of insolvent thrifts were not bailouts nor was the FIRREA a bailout bill. The payments were the direct or indirect honoring of insurance obligations and were not subsidies. Nevertheless, the terms have stuck and will probably remain with us. This is unfortunate because they convey the wrong impressions as to the uses and beneficiaries of the moneys, and they have had harmful consequences. Word usage can shape thoughts and actions. At least part of the reason why the Congress treated the solvent portion of the thrift industry so harshly in the FIRREA legislation is that the press, the public, and the Congress believed that they were somehow "bailing out the thrifts." If, as seems quite likely, the Congress will again have to address the issue of adequate funds for cleaning up the remaining insolvencies, the pervasive use of the term *bailout* will continue to influence its actions in unfortunate ways.

Were the 1988 Transactions Financially Sound?

The FSLIC's placement of 179 thrifts with acquirers during 1988— especially the seventy-five placed in a flurry of thirty-five transactions in December—generated considerable controversy, for the reasons discussed previously. Their cost to the FSLIC was estimated at $27.1 billion. The cost to the U.S. Treasury in reduced future tax collections was estimated at another $5.5 billion. Together, the estimated cost was $32.6 billion.

The sums were clearly large, and the transactions have continued to arouse controversy. The following discussion of their structure and logic and of some evidence on their financial soundness should help clarify the issues.

First, the obligation to honor the insurance promise to depositors required payments by the FSLIC that were roughly equal to the negative net worth "hole" in these deeply insolvent thrifts. Regardless of the disposal method chosen, sizable payments were going to be necessary.

Comparisons are often made between the FSLIC's payments and the acquirer's cash infusion in a transaction. These comparisons are meaningless. The FSLIC's insurance obligation meant that it had to fill in the insolvency "hole," regardless of its depth, caused by the thrifts's *past* actions; the acquirer's infusion was necessary to establish an ownership stake and provide a protective buffer to the FSLIC for the thrift's *future* operations.

Second, for each transaction the FSLIC estimated both the present discounted cost of the transaction and the present discounted cost of liquidating the thrift. In each transaction the FSLIC could not proceed unless the former was less than the latter.[38] For all 179 thrifts these estimated costs of liquidation summed to $37.3 billion. Thus, *the estimated costs of the transactions, even including the forgone tax collections by the Treasury, were $4.7 billion less than the estimated cost of liquidations.* This latter cost implicitly assumed that the FSLIC somehow would have had the $100 billion in immediate cash that would have been necessary to liquidate these institutions.[39]

The FSLIC was supposed to ignore the tax losses to the Treasury in these comparisons and base its decisions only on its own costs. However, in only five of the FSLIC's eighty-six transactions during 1988 did the inclusion of the tax costs to the Treasury raise the estimated overall costs of transaction above the estimated cost of liquidation, and the amounts were small in absolute and/or relative terms. Even in these instances proceeding with the transactions was superior to the alternative for those thrifts. Realistically, the alternative was deferring liquidation until some indefinite time in the future. In the interim, these insolvent thrifts were running large operating losses—typically at a percentage rate that was greater than the interest costs on the FSLIC's borrowings. Thus, the future costs of liquidation would have been appreciably greater. (Equivalently, the FSLIC should have been using a higher discount rate to reflect its opportunity cost in not disposing of these heavy bleeders more rapidly; a higher discount rate would have made liquidation relatively more costly.) Also, proceeding with the transactions would have an important, uncounted benefit as compared with delay: The replacement of an insolvent thrift with a solvent successor would reduce deposit interest costs, not only for the successor thrift but also for other thrifts in the same market area.[40]

Third, the potential tax breaks for acquirers were seen as an extra asset for the FSLIC, which could be used to reduce its costs. The allocations of these tax breaks were explicitly negotiated in each transaction, and the FSLIC usually received a specified percentage of the acquirer's tax benefits; in essence, the FSLIC's payments to the acquirer would be reduced in relation to a specified percentage of the acquirer's reduced tax payments to the Treasury. For all eighty-six transactions in 1988 the estimated reduction in the FSLIC's payments came to slightly more than half of the estimated $5.5 billion ($2.8 billion, or 52 percent) reduction in tax collections by the Treasury. Thus, the FSLIC directly recaptured over half of the tax benefits.

The remaining tax benefits ($2.7 billion) appeared to be retained by the acquirer, in the first instance. In return for granting the retention of those tax benefits to the acquirer, however, the FSLIC could obtain more favorable terms (lower payments) on other portions of the contractual arrangement with the acquirer. If the FSLIC had negotiated perfectly, it would have achieved a $1 reduction in its payments to the acquirer for every $1 of tax benefits retained by the acquirer.

The effectiveness of the FSLIC in negotiating these tax benefits has been tested in a study that examined the determinants of the FSLIC's estimated cost savings (as compared with the estimated costs of liquidation) in the eighty-six placements with acquirers in 1988.[41] The study found that, controlling for other factors,[42] the FSLIC reduced its costs (payments to acquirers) by approximately 85 cents for every dollar of tax benefits generated in these transactions.[43] Thus, though the FSLIC apparently did not negotiate perfectly with acquirers over these tax benefits, it certainly did not squander them and *managed to recapture, directly or indirectly, more than five-sixths of the estimated tax loss to the Treasury.*

Fourth, the stretched-out nature of the payments—through the capital loss coverage and yield maintenance arrangements on covered assets—were a way of deferring the valuation of these difficult-to-value assets and leaving their management in private hands, along with deferring payments. Any arrangement concerning these assets— short of outright transfer to the acquirers at prices that were likely to be unfavorable to the FSLIC—would have involved difficult incentive problems for their management and sale.

Fifth, a major goal of the Bank Board's Southwest Plan was to reduce or eliminate the Texas premium that Texas thrifts were having to pay to retain their deposits.[44] Table 8-5 shows the typical differential (higher interest rates) that thrifts in Texas were paying on newly issued CDs, as compared with the national average of CD rates for the same maturities offered by all thrifts.[45] For Texas insolvents, the differential spiked in June 1987 and remained around 0.75 percent (75 basis points) for the remainder of the year. The differential had fallen substantially by March 1988 (the Southwest Plan was announced in January), rose modestly during the middle of the year, and fell again at the end of 1988. Even solvent Texas thrifts had to pay a substantial premium during 1987; their average premium was substantially reduced during 1988.

The placement of eighty insolvent thrifts with acquirers during 1988 (along with one liquidation and eight "stabilizations"[46]) surely played a role in this reduction in the Texas thrifts' relative deposit costs.[47] Since these insolvent thrifts (and some of the then-solvent thrifts) were wards of the FSLIC awaiting future disposal, these reductions in deposit costs were directly beneficial in reducing the future costs of their disposal.

Sixth, judgments about the effectiveness of FSLIC's negotiations with acquirers and the overall costs of each transaction are difficult to make. (Recall, though, that the estimated cost of each transaction had to be less than the estimated cost of liquidation.) In an important sense, the FSLIC could afford to bargain neither too hard nor too leniently. If its payments were insufficient to fill the negative net worth "hole" adequately, the successor thrifts would likely fail and return to the FSLIC's caseload; the FSLIC would surely be criticized for concluding the transactions with inappropriate acquirers. If its payments were too generous, the FSLIC would be criticized for "leaving too much money on the table" and unduly favoring acquirers. In short, the FSLIC had to meet a "Goldilocks standard" for these

Table 8-5

Differentials (Higher Interest Rates) Paid by Texas Thrifts on Certificates of Deposits, as Compared with the National Average for All FSLIC-Insured Thrifts, 1987–1989

	Insolvent Texas Thrifts[a]			Solvent Texas Thrifts[a]		
	2–3 mo.[b]	3–6 mo.[c]	1–2 yr.[c]	2–3 mo.[b]	3–6 mo.[c]	1–2 yr.[c]
1987						
March	0.41%	0.64%	0.66%	0.33%	0.56%	0.52%
June	1.01	1.02	0.97	0.33	0.72	0.57
September	0.77	0.75	0.76	0.35	0.68	0.58
December	0.73	0.78	0.70	0.27	0.59	0.29
1988						
March	0.27	0.35	0.36	0.19	0.27	0.22
June	0.38	0.47	0.45	0.16	0.34	0.22
September	0.41	0.48	0.48	0.13	0.28	0.22
December	0.31	0.34	0.38	0.26	0.19	0.14
1989						
March	0.65	0.64	0.59	0.06	0.36	0.32
June	0.55	0.56	0.47	0.19	0.33	0.36

[a] Measured on a GAAP basis.

[b] On newly issued CDs in amounts between $80,000 and $100,000.

[c] On newly issued CDs in amounts below $80,000.

Source: FHLBB data.

transactions: Its payments had to be be neither too small nor too large; they had to be "just right."

In this light, some of the FSLIC's transactions of the early 1980s were clearly too thin. By encouraging acquirers to create goodwill assets to fill the negative net worth "holes" of the insolvent thrifts, the FSLIC was not providing the acquirers with sufficient earning assets; some of these acquirers subsequently failed.[48]

In 1988, though, the FSLIC was criticized for lavishing excessively large sums on acquirers. To some extent these criticisms were rooted in the misunderstandings of the necessity for FSLIC "hole-filling" payments, the confusions engendered by the use of the word "assistance," and simply the sizes of the sums being paid. Still, it is quite legitimate to try to ascertain whether the FSLIC's 1988 transactions were "too rich."

At one level, the FSLIC's efforts to attract the interest of as many acquirers as possible to these transactions—in essence, to hold an auction to obtain the least costly bid for each transaction—and the necessity that the estimated cost to the FSLIC of any successful bid had to be less than the estimated cost of liquidation should offer some reassurance on this question. Nevertheless, all of these transactions required detailed negotiations with each successful acquirer, and one should not rule out the possibility that the FSLIC personnel were out-negotiated by the acquirers.

One other point is worth noting: During 1988 these transactions were perceived by actual and would-be acquirers as quite risky. The financial strength of the FSLIC's notes and guarantees, which were significant fractions of the assets acquired, were questioned by the accounting profession early in 1988; these doubts dissuaded many publicly traded companies from making bids.[49] Further, as 1988 proceeded and it became increasingly clear that the future revenues of the FSLIC would be inadequate to deal with all of the costs that it faced and that further Congressional action with respect to funding would be needed, these doubts grew.[50] Also, acquirers were fearful as to what future actions the Congress might take with respect to the thrift industry generally or even with respect to the specific acquisitions. Finally, there were some observers who were questioning the future role of the thrift industry in the financial services markets, which added to the uncertainties faced by acquirers. In sum, acquirers perceived significant risks at the time, and any assessment of these transactions should incorporate these at-the-time perceptions.

One method of evaluating whether these transactions were too rich would be to examine the stock market's reaction, with respect to the stock price of an acquirer (if the acquirer's shares were publicly traded), to the announcement of the acquisition. If the acquirer had achieved an especially good deal—if the FSLIC had left too much money "on the table"—the stock market should have reacted favorably to the "event," and the increase in the acquirer's stock price (corrected for any expected changes due to changes in the overall stock market) should approximate the stock market's guess as to the net gain that the acquirer received in the transaction. One such "event" study examined fifteen transactions during 1988 involving twenty-two insolvent thrifts that were placed with other (publicly traded) thrifts.[51] The study found that the stock market reactions were, on average, positive; however, the aggregate increase in share values for all fifteen transactions summed to only 4.6 percent of the FSLIC's estimated costs for these transactions. This 4.6 percent estimate is probably well within the uncertainty band as to the actual costs of these transactions and within the margin that superior management of the assets of the insolvent thrifts by their acquirers could save for the FSLIC. Thus, this evidence does indicate that the FSLIC may have left a little money "on the table" in these transactions— but probably less than 5 percent.[52]

Seventh, and finally, the FSLIC negotiated considerable flexibility into the transactions. Call provisions were attached to most of the notes issued and to the covered asset arrangements. These notes and covered asset contracts appear to be the most controversial aspect of the transactions, since they involved tax breaks for the acquirers and they involved continuing payments by the FSLIC for up to ten years. If future evaluations should indicate that these notes and contracts were too costly, they can be called in and paid off at that time.

A Summing Up

The FSLIC's 1988 transactions are likely to remain controversial. Their financial magnitudes and complicated structures should not, however, cloud their underlying logic: The FSLIC was obliged to honor its insurance obligations to depositors, and this required the disposal of hundreds of deeply insolvent thrifts. Large sums were going to be necessary under any disposal method. Many complex transactions were negotiated, and improvements may well be possible in the future. But the contracts already contain substantial flexibility to achieve those improvements. In aggregate, the transactions were cost-effective ways of discharging those responsibilities and reducing the ultimate cost to U.S. taxpayers.

Notes

1. The present discounted value of the tax breaks to acquirers were estimated to be an additional $5.5 billion. In addition, the FSLIC "stabilized" eight thrifts in Texas and ten thrifts in Oklahoma. These thrifts were placed into pass-through receiverships and consolidated into five new institutions, in preparation for eventual placements with acquirers. The FSLIC estimated that those placements would cost an additional $7 billion.
2. Those profits, though, did mean that the healthy thrifts could continue to pay insurance premiums to the FSLIC.
3. This Texas premium was not being created by aggressive, risk-taking thrifts that might be trying to expand. Rather, it was the consequence of depositor jitters concerning the financial health of the Texas thrifts and of the FSLIC.
4. In 1987 the FHLB of Dallas had begun a program of collecting moneys from solvent thrifts throughout the country and depositing them (in $100,000 packages) in insolvent thrifts, in order to increase the supply of relatively low cost deposits to these insolvent thrifts. At its peak this program had collected and deposited approximately $2.7 billion.
5. In October 1988 the Congress renewed the tax breaks for only six months and encumbered them with reduced coverage and effectiveness.
6. On a discounted present value basis.
7. To be consistent with the estimates of the FSLIC's disposal costs, these estimated costs of forgone tax collections were also computed on a present discounted value basis. These estimates required judgments as to the likely future profitability of the acquirers and their continued ability to use the tax breaks.
8. Eighty were placed with acquirers; one was liquidated.
9. Technically, the liquidation of assets is done through the appointment of a liquidating receiver.
10. This $60 would have to be net of any operating expenses of the liquidating receivership.
11. In 1989 and 1990 this need for initially larger sums of cash for liquidations came to be known as the problem of "working capital" for the FSLIC's successor, the Resolution Trust Corporation.
12. The following discussion explains why a payment that was somewhat less than $32 could be adequate for an acquirer.

13. A liquidation of an insolvent thrift through a transfer of only the deposits to another thrift or a bank, which would require all cash at the time that the deposits were transferred, and a placement of the "whole" insolvent thrift with an acquirer, which would require cash only to fill the negative net worth "hole" of the insolvent thrift, represent two extreme points along a continuum of combinations of cash and the insolvent thrift's assets that the FSLIC could transfer to the acquirer. Intermediate combinations were possible, in which only some (but not all) of the insolvent thrift's assets were transferred to the acquirer, with the FSLIC's keeping the remaining assets and giving larger amounts of cash (or notes) to the acquirer instead.

14. For the FSLIC during 1988, tax breaks were a third reason. By law the FSLIC was supposed to choose the option that was least costly to it alone; tax losses to the Treasury were supposed to be ignored. In practice, in only five of the eighty-six transactions in 1988 did the inclusion of the Treasury's tax losses cause the estimated all-inclusive government cost of placing an insolvent thrift with an acquirer to exceed the estimated costs of liquidation. In all five instances the extra costs (above estimated liquidation costs) were small in absolute and/or percentage terms.

15. This was clearly the case for two large liquidations in June 1988: North American Savings of Santa Ana, California, and American Diversified Savings of Lodi, California.

16. The FSLIC could and did contract with private real estate management and disposal companies, in order to bring private sector expertise into the picture. The basic cost-reimbursement terms and incentive structures of those contracts, however, were always a difficult problem; and ultimate decisional authority would still rest with government employees. Also, the FADA did not turn out to be the solution to the FSLIC's problems of asset management and disposal; see Note 28, Chapter 7.

17. The Bank Holding Company Acts of 1956 and 1970, and their interpretations by the Federal Reserve, prohibited companies engaged in nonfinancial commerce or industry or in insurance from owning a bank; the Glass-Steagall Act of 1933 prohibited investment banking companies from owning a commercial bank.

18. Since the FSLIC was constrained by limits on the available amounts of cash, the inability of the FSLIC to place a thrift with an acquirer meant that the thrift would be "warehoused" until sufficient cash was available for the liquidation.

19. The FSLIC had to use its best guess as to the timing of the sale or disposal of various types of assets under either scenario and the likely price that the asset would bring at that time; these values were then discounted to the present (i.e., the date of the decision concerning the transaction).

20. The interest rates on the notes were typically adjusted every three months and were linked to the average interest rates on deposits of all thrifts in that thrift's state. For example, notes issued to acquirers of

insolvent Texas thrifts usually carried interest rates that were 0.40 percent (40 basis points) above the average deposit costs of all thrifts in Texas.

21. These difficulties are not a legitimate argument against the need for market value accounting, which will be advocated in Chapter 11, for at least two reasons. First, these assets were exceptional; they were the consequences of the debacle described in Chapter 6. Second, had market value accounting been applied consistently, even to these assets, better approximations of their market values would have been readily available to the FSLIC and to the acquirers.

22. In these yield maintenance arrangements the FSLIC provided payments, in addition to any income that the asset itself might yield, so as to achieve an acceptable rate of return on these assets. The target rate of return, like the interest on the FSLIC notes, was linked to the average interest costs on deposits in the state. The margin over those costs was substantially larger, however, to allow for the greater costs of managing these assets. But the margin was scheduled to decline over the ten years of the contract, so as to encourage the acquirer to dispose of the assets rather than just hold them and receive the yield maintenance payments. In Texas, for example, an acquirer of an insolvent thrift might have a contract promising yield maintenance on specified assets that began, say, at 2.50 percent (250 basis points) above the average cost of deposits in Texas and that then declined in annual or biannual jumps to, say, a 1.75 percent (175 basis point) margin by the tenth year.

23. The FSLIC would thereby have been accused of "giving away" the (subsequently valuable) Texas assets.

24. In the language of formal economics, the FSLIC faced a difficult "agent-principal" problem, regardless of how it chose to handle these assets.

25. As of early 1990, the experience of the Resolution Trust Corporation in dealing with these difficult assets indicates that it too has found no easy solution to these agent-principal problems.

26. In the case of the placement of American Savings of Stockton, California, with a group headed by Robert Bass, the FSLIC retained warrants equal to 30 percent of the stock of the new thrifts.

27. Because of legal prohibitions the FSLIC could not own the stock of a thrift directly.

28. The warrants position of the FSLIC would also reduce the incentive for risk-taking by the acquirer, since the acquirer would receive less of the benefits of the risk-taking; for a similar idea, see Bernheim (1988).

29. The FSLIC note retained its special tax status if it was issued in connection with the placement of an insolvent thrift with an acquirer and if it continued to be held as an asset by the successor thrift.

30. The origins of the term are hard to trace. It probably arose in the context of improper analogies with earlier government aid to Lockheed

and Chrysler. Since government money was involved in these thrift transactions, there was a surface similarity. The difference, of course, is that the FSLIC's payments were the honoring of a previously established insurance guarantee rather than an ad hoc arrangement to keep a company in operation.

31. In the event of violations of fiduciary obligations or criminal transgressions, however, directors and officers are liable for civil or criminal penalties.

32. In a very few instances, where the insolvency was thought to be largely the consequence of local economic conditions and there were no suspicions of irregular behavior by senior managers, acquirers kept them on, especially if they were considered important for retaining the thrift's ties with its community.

33. Uninsured depositors would share proportionately in the proceeds of the liquidation, alongside other creditors and the FSLIC.

34. Only in the case of American Savings of Stockton, California, was an explicit promise extended to uninsured depositors.

35. It is worth noting, however, that thrifts (and banks) pay insurance premiums on the uninsured amounts.

36. This has been less true for FDIC transactions with respect to insolvent commercial banks. In these cases the amounts of nominally uninsured deposits have been large, and the FDIC has often chosen to protect them.

37. The FDIC uses similar phrases for its payments.

38. One study (Kormendi et al., 1989) criticized the FSLIC's cost comparisons, arguing that a lower discount rate and the inclusion of the franchise value that could be received from the transfer of core deposits (i.e., the stable, loyal depositor base) would have lowered the estimated costs of liquidation. On the other hand, liquidation was not a realistic, immediate alternative. Further, given the large and rising operating losses by these insolvent thrifts, an even higher discount rate could have been justified. An important externality or spillover effect of placements with acquirers (as compared with the realistic alternative of doing nothing)—the lower deposit interest costs that would be enjoyed by other thrifts in the same market area—was being ignored. And the FSLIC's standard calculation for estimating the costs of liquidation was underestimating the recent actual costs of the operations of the liquidating receiverships by amounts that more than offset the neglect of the core deposit franchise value.

39. The experience of the Resolution Trust Corporation after August 1989 has indicated that, even with more resources, rapid liquidations were not likely.

40. See Note 26, Chapter 7.

41. See Barth, Bartholomew, and Elmer (1989).

42. Principal among these were the size of the core deposits, the amount of mortgage servicing, the branch structure, the type of acquirer, and the extent of "bad" assets. The study found that the Southwest Plan (Texas) transactions were not significantly different from the remaining transactions in 1988 in terms of the cost savings (as compared to liquidation).

43. This differs from the conclusions reached by Kormendi et al. (1989). Their study, however, neglected a number of the important variables included by Barth, Bartholomew, and Elmer (1989), especially the size of core deposits.

44. See Note 3.

45. The focus on newly issued CDs of specific maturities avoids the muddying of comparisons that involve just "the cost of funds." The latter concept would include the effects of existing (previously issued) CDs and of changes in the maturity mix even of newly issued CDs.

46. See Note 1.

47. Also, the Dallas FHLB's program of channeling moneys from solvent thrifts into the deposits of Texas insolvents surely played a role; see Note 4.

48. The failure of Empire of America Savings of Buffalo, New York, in early 1990 was notable example.

49. The executives of publicly traded companies feared that they would have to include qualifying or explanatory footnotes to their financial statements if they acquired FSLIC notes or guarantees and that these footnotes would frighten their stockholders and creditors.

50. Even some Congressmen and Senators raised the possibility, which was (fortunately) subsequently not pursued, that the Congress might explicitly repudiate the FSLIC's notes.

51. See Cole, Eisenbeis, and McKenzie (1990). Their study covered a larger group of forty-four transactions (involving fifty-eight insolvent thrifts) over the period 1980-1988. Their results are similar to those of Balbirer, Jud, and Lindahl (1989).

52. Unfortunately, the study covers only acquisitions by publicly traded thrifts and does not include many of the more controversial acquisitions, such as those by Robert Bass (American Savings of Stockton, California), Ronald Perelman (five large thrifts in Texas, including First Texas Savings of Houston and Gibraltar Savings of Dallas, Texas), and William Simon (Bell Savings of San Mateo, California). For the conclusions of the study not to apply to the other transactions of 1988, one would have to assume that the nonthrift acquirers somehow were appreciably and systematically better negotiators than the thrift acquirers; for other discussions of the Southwest Plan transactions, see Horvitz (1989a; 1989b) and Klinkerman (1990).

CHAPTER NINE

The FIRREA, 1989

As the Federal Home Loan Bank Board was feverishly concluding its transactions with acquirers of insolvent thrifts in late 1988, the newly elected but not-yet-inaugerated Bush administration was gathering information for a new assault on the problems of the insolvent thrifts. As 1988 progressed the depth and extent of the insolvent portion of the thrift industry became increasingly clear. Despite the FSLIC's disposal of 205 insolvent thrifts (and "stabilization" of another 18), the industry ended the year with 243 thrifts (with $74.3 billion in assets) that were insolvent according to regulatory accounting principles. Twice that number (with four times the assets) were insolvent on a tangible net worth basis. One-third of the industry was unprofitable. That third of the industry experienced $6.8 billion in operating losses and $12.2 billion in nonoperating losses (writedowns), swamping the $5.6 billion in after-tax profits earned by the profitable two-thirds. The "chickens" of the 1983-1985 period were still "coming home to roost."

In this light, it was increasingly apparent that the Competitive Equality Banking Act of 1987 had been inadequate. The CEBA's borrowing mechanism had provided little in the way of new resources; it was primarily a means of telescoping future insurance premiums into the present. With the continuing deterioration of the Southwest real estate markets and the unabated sliding of thrifts into insolvency, the FSLIC's net resources—primarily, its future insurance premium revenues—were increasingly seen as inadequate to handle its obligations to the insured depositors in these insolvent thrifts. A new plan was necessary.

On February 6, 1989, President Bush announced his new program, which he would subsequently send to the Congress for legislative enactment. The program called for an additional $50 billion in borrowing authority to clean up the remaining insolvent thrifts. General revenues would explicitly be used to cover a large fraction of the necessary costs; the healthy part of the thrift industry would be taxed to cover the remainder. The Bank Board and the FSLIC would be abolished; the regulatory, deposit insurance, and lending functions of the agency were to be transferred to other government agencies. In the interim, the president asked the FSLIC to cease virtually all disposal transactions and to transfer control of all insolvent thrifts to the FDIC, which would succeed the FSLIC as the deposit insurer for all thrifts. Higher net worth standards for thrifts would be mandated. And extra moneys would be allocated to the U.S. Department of Justice for stepped-up criminal prosecutions.

The Bush administration spent the next month developing the specifics of the proposal before forwarding a detailed legislative package to the Congress in early March. Despite the recognized urgency of action—the FSLIC was no longer disposing of the insolvents, and thrifts' losses were mounting—the Congress took five months to pass the legislation.[1] On August 9, 1989, President Bush signed into law the Financial Institutions Reform, Recovery, and Enforcement Act (FIRREA) of 1989.

The FIRREA: The Basic Structure

The final form of the FIRREA was similar to the Bush administration's proposals in most ways, but the Congress added elements of its own. The legislation embodied the most sweeping changes in thrift regulation and deposit insurance since the Home Owners' Loan Act of 1933 and the National Housing Act of 1934. The printed version of the law runs to 381 pages. Briefly, the major features of the legislation are as follows:

1. *Funding.* The FIRREA authorized $50 billion of additional borrowing to deal with the costs of disposing of the remaining insolvent thrifts. A dispute between the Congress and the Bush administration as to the budgetary accounting for this borrowing led to a compromise in the legislation: The first $20 billion was borrowed through general Treasury ("on-budget") bond financing before September 30, 1989.[2] The remaining $30 billion will be borrowed ("off-budget") through a new financing entity, the Resolution Funding Corporation (REFCORP). The principal amounts of the bonds issued by the REFCORP will be guaranteed by zero coupon Treasury bonds, as was true of the FICO bonds; however, the interest payments will be guaranteed by the Treasury.

General Treasury revenues will be used to cover a high fraction (at least three-quarters) of the costs. The remainder is expected to come from the healthy sector of the thrift industry through two routes: higher insurance premiums and the taxing of the net worth and future profits of the Federal Home Loan Banks (which are owned by the thrift industry).

The schedule of future deposit insurance premiums for thrifts (and banks) mandated by FIRREA is provided in Table 9-1. As can be seen, thrifts' premiums will rise from current levels during 1991-1993 and then fall subsequently, eventually reaching par with commercial banks' premiums in 1998.[3] Though the FIRREA also raised commercial banks' premiums, it established two separate insurance funds within the FDIC, and the bank premium revenues will be solely devoted to the bank fund and not used to defray the costs of cleaning up the problems of the insolvent thrifts. With a substantial differential in insurance premiums, many thrifts would be tempted to switch their charter and qualify for the banks' insurance fund. Since the intent of FIRREA is to tax the thrifts, it imposes a five-year moratorium (with

Table 9-1
Annual Deposit Insurance Premiums Mandated by the
FIRREA, 1989–1998

| | Amounts per $100 of Deposits | |
	Thrifts[a]	Commercial Banks[b]
1989	20.83 cents	8.33 cents
1990	20.83	12.0
1991–1993	23.0	15.0[c]
1994–1997[d]	18.0	15.0
1998 and after	15.0	15.0

[a] Thrifts that were previously insured by the FSLIC prior to August 1989; the premium rates apply to the deposits insured by the FDIC's Savings Association Insurance Fund.

[b] Includes thrifts (MSBs) that were insured by the FDIC prior to August 1989; the premium rates apply to the deposits insured by the FDIC's Bank Insurance Fund.

[c] In August 1990 the FDIC proposed that BIF premiums for 1991 would be 19.5 cents.

[d] Beginning in 1995, the FDIC can increase premiums by 7.5 cents per year, up to a maximum level of 32.5 cents, for either fund, if the reserves in that fund are deemed insufficient; the FDIC can increase BIF premiums even earlier, beginning in 1990, but more stringent criteria apply through 1994.

Source: FIRREA.

some exceptions) on such attempts and also requires that exit and entry fees be paid for any such switch.

The FHLBs are taxed in two ways. The FIRREA requires the FHLBs to contribute slightly more than $2 billion of their net worths (retained earnings) in 1989 and 1990 to the REFCORP.[4] Further, the FHLBs are required to contribute $300 million annually from their profits to the REFCORP. (In addition, the FHLBs are required to contribute at least $50 million annually to affordable housing programs through 1993, $75 million in 1994, and $100 million in 1995 and subsequent years.)

2. *Organizational Structure.* The FIRREA abolished the Bank Board and the FSLIC and scattered their functions in the following directions:

a. *OTS.* The thrift chartering and primary regulatory functions were reconstituted in a newly established bureau within the Treasury, the Office of Thrift Supervision.[5]

b. *FDIC-SAIF.* Thrift deposit insurance was transferred to the Federal Deposit Insurance Corporation, which had previously insured the deposits in commercial banks and mutual savings banks. As insurer of thrifts' deposits, the FDIC also acquired secondary regulatory responsibilities. Two separate insurance funds were created within the FDIC: the Savings Association Insurance Fund, insuring the deposits of the former FSLIC-insured thrifts; and the Bank Insurance Fund, for the commercial banks and MSBs. As noted earlier, there is a five-year moratorium on healthy thrifts' switching out of the SAIF and into the BIF.

c. *RTC.* The responsibility for disposing of the insolvent thrifts and their assets, formerly handled by the FSLIC, was assigned to a new entity, the Resolution Trust Corporation (which would be funded by the REFCORP). The RTC, however, is managed by the FDIC and has, for practical purposes, become another branch of the FDIC. The FIRREA further established a separate entity, the RTC Oversight Board, to set policy for the RTC and review its actions. The RTC Oversight Board has five board members (the Secretary of the Treasury, the Secretary of Housing and Urban Development, the Chairman of the Federal Reserve System, and two appointed members[6]) and a small staff.

d. *FHFB.* The FIRREA assigned the Bank Board's oversight of the FHLBs' lending activities to a new entity, the Federal Housing Finance Board. The FHFB has five members (the Secretary of HUD and four appointed members[7]) and a staff.

3. *Regulation.* The FIRREA tightens thrift regulation in a large number of ways. The following are some of the more important changes:

a. *Qualified thrift lender test.* Thrifts are expected to devote 70 percent of their assets to housing-related investments. This qualified thrift lender (QTL) test is higher than the 60 percent QTL requirement that had been imposed by the CEBA and is more stringently applied.

b. *Net worth standards.* The FIRREA mandated net worth standards that are no less stringent than the standards that apply to national banks (which have been higher than those that have applied to thrifts). The Congress also specified a phase-out transition rule for allowing limited amounts of goodwill to be included as an asset in the net worth calculations. Following the FIRREA's mandate, the OTS in November 1989 established one set of interim net worth standards for 1990, a higher set of interim standards for 1991-1992, and a yet higher final set ("fully phased-in") for 1993 and after.

c. *State-chartered thrifts.* The FIRREA restricts the activities permitted for state-chartered thrifts to those permitted for federally chartered thrifts, regardless of state authorizations. State chartered thrifts that meet their fully phased-in net worth requirements, however, may engage in activities that are not authorized for federally chartered thrifts (or engage in activities to a greater extent, if the activity by federals are allowed but limited), if the FDIC finds that the activities are not unduly risky.[8]

d. *Federally chartered thrifts.* The FIRREA imposes new restrictions on federally chartered thrifts (and thus for all thrifts, through the logic of the previous section). Thrifts must meet the standards that apply to national banks for loans to a single borrower; this provision effectively reduces the maximum size loan to a single borrower from 100 percent of a thrift's net worth to 15 percent. Thrifts cannot acquire any new "junk bonds" and must divest their existing holdings by July 1, 1994. In addition, the FIRREA lowered thrifts' maximum permitted holdings of commercial real estate mortgages from the 40 percent of assets permitted by the Garn–St Germain Act to a new level of four times net worth.

e. *Enforcement.* The FIRREA expands the role and increases the amounts of civil and criminal money penalties for legal and regulatory violations. The ability of the OTS to issue cease and desist orders and to remove directors and officers of thrifts is expanded. The FIRREA gives the FDIC new powers to terminate or suspend the deposit insurance of a thrift (or a commercial bank) at relatively short notice. And the Act authorizes the expenditure of an additional $75 million for the U.S. Department of Justice to prosecute criminal activities committed in relation to thrifts or banks.

A Critique of the FIRREA

1. *The anger and its consequences.* In an important sense, the FIRREA was an Act of anger. The Congress and the Bush administration were angry over the necessity to spend large sums to clean up the problems of the insolvent thrifts. They were angry at "the thrift industry," whose members were directly responsible for the insolvencies. The Congress believed it was "bailing out the thrifts," rather than being asked to satisfy its obligations on an insurance arrangement that earlier Congresses had created and whose poor administration earlier Congresses had either encouraged or willfully ignored. The Congress's anger was also focused on the industry's past lobbying efforts, especially the industry's efforts in 1987 to reduce the CEBA funding to $5 billion and the efforts of individual thrifts to have individual Congressmen and Senators intercede on their behalf in specific regulatory matters.[9]

The Congress's anger extended to the owners and managers of healthy, solvent thrifts as well as to the operators of the insolvents, even though the former had not extended the risky loans or engaged in fraudulent activities. Their major failing had been the exercise of their First Amendment rights to lobby for measures that were not in the larger national interest; theirs was not a noble cause or effort, but they were far from alone in this respect on Capitol Hill. This did not matter. In an important sense the Congress could not distinguish between the operators of the solvents and the insolvents. As late as 1987 the Congress had thought that all thrift operators were Jimmy Stewart in *It's a Wonderful Life;* in 1989 the Congress thought they all were Warren Beatty in *Bonnie and Clyde.* Since the Congress thought it was somehow bailing out the thrifts, it therefore expected "the thrifts" to bear part of the costs and to endure the appropriate punishment. The fact that the remaining *healthy* thrifts would be the ones to bear the costs and punitive consequences was little noticed or understood. The Bush administration did not try to dissuade the Congress of these views.

The Congress and the Bush administration were also angry at the Federal Home Loan Bank Board. The anger encompassed the Bank

Board's earlier lax regulation as the proximate cause of the insolvencies, the agency's historical coziness with and championing of the (now pariah) thrift industry, the agency's 1988 "bailout" transactions, and the agency's underestimates of the aggregate size of the insolvency problem during 1988. Though the Bank Board since 1979 had gone through five chairmen, eight board members, about as many directors of the FSLIC, and numerous turnovers of its senior managerial staff, the anger was personalized: "The Bank Board" was to be punished for "its" sins. Forgotten in this passion was the role of earlier Congresses in encouraging laxity and then turning a deaf ear to its consequences—as recently as August 1987 in the passage of the CEBA. Also neglected was the recognition of the Bank Board's improvements in regulatory procedures and personnel since 1985 that were discussed in Chapter 7. An understanding of the necessity for the 1988 transactions, and their large sums, was wholly lacking. Again, the Bush administration did little to educate the Congress on these points.

This anger toward the thrift industry and toward the Bank Board manifested itself in a number of the provisions of the FIRREA, with deleterious consequences for all concerned. The healthy remainder of the thrift industry was clearly being punished. They were asked to pay higher insurance premiums for 1991–1993 and a continued differential vis-à-vis banks until 1998, as is shown in Table 9-1. The higher premiums would mean lower profits for the remaining healthy thrifts. Second, the FIRREA taxed and reduced their stakes in and claims to the profits and dividends of the FHLBs, again lowering their profits and their net worths. Third, their operating flexibility—even for prudently managed, well-capitalized thrifts—was reduced by the higher QTL test, the reduction in maximum loan size, and the reduced asset diversification authority.

These acts of anger toward the thrift industry revealed a clear strain of schizophrenia running through the Congress. The FIRREA mandated higher net worth levels for thrifts; yet the taxes and restrictions, and the possibility of even more punishments in the future, made it much more difficult for the remaining thrifts to earn the profits and raise the capital necessary to attain those levels. Further, many of those who voted for the FIRREA claimed that it was inevitable and desirable for thrifts to become more like commercial banks—yet they voted for the higher QTL that would force thrifts once again to become more specialized on residential mortgage finance.

Ironically, these punishing taxes and restrictions have had consequences that extend far beyond the healthy thrifts. They have had a rebound effect that has increased *the public's costs* of cleaning up the insolvents. First, some marginal thrifts that might have survived and raised fresh capital to bolster their net worths have been given an extra shove toward insolvency. Second, in the early months of its disposal efforts the RTC experienced little success in finding acquirers for the hundreds of thrifts in its caseload. This delay has added to the costs of the cleanup. At least part of the reason for the RTC's difficulties has been that potential acquirers are wary of acquiring a thrift when their insurance premiums will be higher (and could possibly go even higher[10]), their dividends from the FHLBs are reduced, their operating flexibility is reduced, *and the Congress is likely to re-address the problem in 1991 and may well decide to tax and punish "the thrifts" even further.*

With respect to the Bank Board, the punishment was straightforward: abolish the agency and distribute its powers to other agencies. The FIRREA specifically instructed the RTC to re-examine the FSLIC's 1988 transactions, to review all means of reducing the costs of those transactions, and to invoke all legal means of modifying, renegotiating, or restructuring those transactions. The clear sense of the legislation was that the Bank Board as an agency was incompetent (or worse) and that the agency should be ignominiously expunged from the rolls of federal government agencies.

The placement of the OTS within the Treasury was considered quite important since the Congress believed that the OTS would thereby be less susceptible to lobbying from the thrift industry or to pressures from individual Congressmen and Senators and that the Treasury would be a beneficial guiding influence on the OTS. Simultaneously, however, many in the Congress fretted that the FDIC was becoming less independent and more susceptible to the influence of the Treasury. Once again, schizophrenia and the absence of a consistent vision of the proper structure of government were present in the Congress.

This punishment also has had a rebound effect that has increased the costs of the cleanup. The sharp halt in FSLIC disposals of insolvents in February 1989, in response to President Bush's request, meant that for six months no thrifts were liquidated and only one small thrift (with $55 million in assets) was placed with an acquirer. After August 9 a new agency, the RTC (staffed by the FDIC), had to learn the intricacies of insolvent thrifts, their balance sheets, and their shoddy assets, as well as learning about the disposals of these thrifts and the liquidations of their assets. This learning took time, and in the interim relatively few disposals took place. By December 31, 1989, the RTC had disposed of only thirty-seven thrifts, with $9.5 billion in assets. In the meantime, by the end of 1989 the number of RAP-insolvents had risen to 281, with $93 billion in assets. Operating losses in these insolvents, which only added to their eventual costs of disposal, continued to mount.

In sum, the anger expressed in the FIRREA has had its costs. Healthy thrifts will continue to suffer, and more will falter and slide into insolvency. An effective agency was abolished, and a new agency that required learning time replaced it. The momentum of the disposal of insolvents was slowed. Delay has been costly.

2. *The conceptual failings.* There are some beneficial aspects to the FIRREA. But many of its provisions are conceptually flawed.

a. *Borrowings.* Providing more money, to continue the disposals of the insolvent thrifts, was vital. This was the most important aspect of the FIRREA. But limiting the borrowings to $50 billion was a mistake. Instead, the Bush administration should have educated the Congress to the reality that the moneys spent to clean up the insolvent thrifts are the honoring of the U.S. government's deposit insurance obligations and that therefore whatever moneys are required to meet those obligations will have to be spent. The $50 billion limitation encouraged the Congress to believe that this spending was somehow discretionary.[11]

The FIRREA does not anywhere directly state that insured deposits in banks and thrifts are backed by the full faith and credit of the U.S. government. It does so indirectly, however, by requiring thrifts (and permitting banks[12]) to display a sign (logotype) that states that "insured deposits are backed by the full faith and credit of the United States Government." This requirement, and the explicit language of the requirement in the FIRREA, presumably has committed the Congress legally in this respect.[13] This is an important and beneficial step. Insured depositors in the BIF and SAIF should never again have to worry if their deposits are safe, and thrifts insured by the SAIF should not have to pay an interest premium to depositors because of the latter's worries about the strength of the fund.[14] Still, it is not reassuring that the Congress treated the authorization of the $50 billion of borrowing to finance the RTC's disposals of the insolvents as if this spending were discretionary. Also, the FIRREA's requirement that SAIF-insured thrifts use one logo, while BIF-insured banks are permitted to use another logo, is at odds with any goal of reassuring all insured depositors in both the SAIF and the BIF.[15]

Second, the REFCORP funding mechanism (like the FICO before it) involves the issuing of $30 billion in long-term bonds that will not be "full faith and credit" borrowings of the U.S. Treasury. As a consequence the borrowings are likely to carry a higher interest rate—about 0.25 percent (25 basis points), or about $75 million per year over thirty to forty years. This is a needless cost that could have been avoided if normal Treasury financing had been used.[16]

Third, in late 1989 and early 1990 the RTC confronted one of the points discussed in Chapter 8: Liquidations, transfers of deposits, or placements that allow acquirers to take only a small fraction of the assets of the insolvent thrift, which have been the RTC's primary disposal method,[17] require extensive amounts of up-front cash, while the replenishment of some of that cash from the sales of assets takes much longer. The $50 billion in borrowing was, at best, supposed to cover the net costs of disposing of the insolvents; it could not possibly provide a sufficient amount of immediate cash for large-scale liquidations. (The 281 RAP-insolvents that were in the RTC's caseload at the end of 1989 had $90 billion in deposits.) The RTC discovered that it needed "working capital": additional sums to finance the liquidations of insolvent thrifts, while waiting for the assets of these thrifts to be sold. Only at the end of February 1990 did the RTC, the Bush administration, and the Congress reach an understanding that would give the RTC adequate working capital.[18] Again, delay was costly.

b. *Thrift insurance premiums.* As noted earlier, part of the punishment meted out by the FIRREA was to force the remaining healthy thrifts to pay higher deposit insurance premiums in 1991–1993. These higher premiums are also intended to force the healthy thrifts to pay for a part of the costs of cleaning up the insolvents. Though the differential between the thrifts' premiums and the premiums paid by banks will actually narrow, thrifts will be at a greater cost disadvantage vis-à-vis other competitors—for example, mutual funds on the deposit side and mortgage bankers on the lending side. Only in 1994 will thrift premiums be below their level of 1989.

The FIRREA, however, also authorizes the FDIC to increase SAIF premiums, beginning in 1995, if the reserves in the fund are not adequate. SAIF premiums can be raised by 7.5 cents per $100 of deposits per year, to a maximum annual level of 32.5 cents per $100 of deposits. These increases could again harm healthy thrifts and put them at a competitive disadvantage; and the possibility that these increases might occur has surely been a factor in discouraging potential acquirers for the RTC's caseload of insolvent thrifts.

The insurance premium schedule established by the FIRREA continues a fifty-six-year history of flat-rate premiums and of Congressional focus on the premiums primarily as a revenue-gathering mechanism and unwillingness to treat deposit insurance as an *insurance* system. As will be argued in Chapters 10 and 11, an insurance orientation would naturally lead to the realization that insured depository institutions should pay premiums that are actuarially based, that are not geared to revenue gathering, and that are not flat rates but rather are variable among the individual depositories depending on the risks that they pose to the insurance fund.[19]

c. *Taxing the FHLBs.* Another part of the FIRREA's punishment and revenue raising effort was to tax the FHLBs' net worths and future dividends. Again, this tax only harms the healthy thrifts, and the prospects of a Congressional repeat of these actions in the future surely discourages potential acquirers of the RTC's insolvents.

d. *Structural Changes.* The FIRREA's punishment for the Federal Home Loan Bank Board was to abolish it and scatter its functions. The general arguments as to why this was a mistake have already been presented; there are more specific reasons as well.

The Bank Board had encompassed the roles of both safety-and-soundness regulator and deposit insurer for thrifts. For reasons that were never clarified, the supporters of the FIRREA came to believe that this combination of regulator and insurer involved a "conflict of interest" and that the two entities therefore needed to be separated. This was one of the justifications offered for sending the thrift insurance function to the FDIC, while creating the OTS within the Treasury to house the thrift regulatory functions.

This notion of conflict of interest is fundamentally incorrect. The regulatory function establishes the set of rules that, if done properly, protects the insurance fund through limitations on risk-taking by the insured institution. The combining of the two roles within one agency meant that the regulator and insurer were automatically in consonance, not in conflict.[20]

The FIRREA implicitly recognized this consonance by giving secondary or back-up regulatory powers vis-à-vis thrifts to the FDIC, along with the insurance function. Thus, rather than separating the regulator and the insurer, the FIRREA in reality created a regulator (OTS) *and* a regulator-insurer (FDIC). This may have some beneficial effects in protecting the insurance funds by creating multiple entities that can establish rules, monitor thrifts, and thereby discourage excessive risk-taking and errant behavior. This has been the pattern of bank regulation.[21] But, it also has its costs: The actions of the primary regulator are no longer necessarily in consonance with the insurer; extra coordination between regulator and insurer is required; and the personnel for the second regulatory function must be paid.

The wave of thrift insolvencies, and their huge costs, occurred because of the forces, circumstances, and actions described in Chapters 5 and 6, not because of any conflict of interest between the Bank Board's role as thrift regulator and its role as insurer. The FIRREA should not have attempted to separate these roles.

At worst, the potential conflict of interest within the Bank Board was between its role as lender to (and promoter of) the thrift industry through the FHLBs and its role as regulator-insurer of those same thrifts. As noted in Chapters 4 and 7, this potential conflict was handled by creating "Chinese walls" within the FHLBs to separate their lending and regulatory functions. Though this procedure did work satisfactorily, the continued role of the FHLBs as both lender and regulator did create appearance problems and could have justified *that* separation (i.e., the creation of the FHFB).

The movement of the thrift insurance function into the FDIC accomplished little of a tangible nature. The separate logos permitted for the BIF and the SAIF even undercuts any uniform assurance for depositors that might have been achieved. And, with the disappearance of the FSLIC, an important source of diversity of approach and of differing ideas on deposit insurance has been lost.[22] Uniformity of approach can be stultifying. The tradition of the dual banking system is one that encourages diversity. The abolition of a separate FSLIC is a step backward.

In sum, the FIRREA's abolition of the Federal Home Loan Bank Board was a mistake—an act of anger that was based on misperceptions and misunderstandings of the causes of the debacle, of the strengthening of regulation that the Bank Board had achieved between 1985 and 1988, and of the nature of the 1988 transactions. The attempt to separate regulator and insurer was misguided, as was the abolition of the FSLIC. At most, only the separation of the lending function from the regulator, through the creation of the FHFB, could be justified.

e. *Net worth standards.* It is encouraging that the Congress now understands the concept of net worth as a protection for the deposit insurer; but the FIRREA's specific directives were not helpful. First, by directing the OTS to establish specific standards and by specifying details such as the level and type of goodwill to be allowed as assets, the FIRREA has needlessly reduced flexibility and involved the Congress in details that are best left to the regulatory process. In December 1988, before FIRREA was proposed by the Bush administration, the Bank Board had proposed a set of risk-based net worth standards for thrifts that were comparable to the standards set for banks, and the agency was well on its way to promulgating final regulations prior to the passage of FIRREA. Second, by forcing the OTS to issue risk-based net worth standards that were comparable to those of the OCC by November 1989, the FIRREA forced the OTS temporarily to exclude interest rate risk from the risk elements of the net worth standards. Though this lapse is likely to be corrected within a year or two, the absence at any time of incentives to reduce interest rate risk is unfortunate. Third, the FIRREA's endorsement of higher net worth standards remains embedded within the context of the existing (GAAP) accounting framework, rather than within a market value accounting framework; however, only the latter can offer net worth measurements that approximate the real protection for the deposit insurer.[23]

f. *Higher QTL test.* The FIRREA's mandate of a higher QTL test needlessly restricts the flexibility and profitability of well-managed thrifts that can prudently diversify. This is directly contradictory to the goal of those who want thrifts to become more like commercial banks.

The higher QTL test was partly intended as punishment and was partly the result of Congressional misinterpretation of the history of the 1983–1985 period. The Congress saw that those thrifts that had "stuck to their knitting" had been less likely to get into trouble; therefore, it decided that all thrifts should be required to stick to that knitting.

As Chapters 5 and 6 demonstrated, however, this view is an oversimplification and a misreading of that history. It ignores the crucial combination of the opportunities-capabilities-incentives for risk-taking that the economic environment and deregulation (compounded by policy errors) had created. In this light, as will be argued in Chapter 11, it is vital to focus the incentives and the information system in the right direction and then provide a regulatory framework that allows and encourages sensible diversification or specialization rather than one that *forces* costly and excessively narrow specialization.

g. *Other regulatory restrictions.* Here, as with the higher QTL test, the FIRREA needlessly reduced the flexibility and profitability of well-managed thrifts. Additionally, as with the net worth standards, it involved the Congress in details that were best left to the regulatory process. The Bank Board had won a major legal victory a year earlier in a challenge to its regulations that limited the percentage of assets that state-chartered thrifts could devote to direct equity investments.[24] This victory established its ability to use safety-and-soundness federal regulation to override excessively permissive state regulation that endangered the FSLIC insurance fund. Further legislation was not needed.

h. *Enforcement.* The FIRREA's increases in civil and criminal penalties, and the stiffening of enforcement powers, were generally sensible. Also, the FIRREA added new grounds for the appointment of a conservator or receiver, including prospective losses or other actions by a bank or thrift that could lead to insolvency. These should be useful additions to the previous grounds, but it is unfortunate that the FIRREA did not specifically strengthen or clarify the existing grounds of asset dissipation or operating in an unsafe or unsound condition. Finally, the FIRREA's strengthening of the FDIC's ability formally to suspend or terminate a bank's or thrift's insurance at short notice is a welcome measure, though its practical value may be limited. The formal termination of insurance has been a rare procedure, and the FIRREA's provisions may not change that tendency.

3. *The operational shortcomings.* At President Bush's request, the FSLIC ceased disposing of insolvent thrifts in early February 1989. Over the next six months, until the passage of the FIRREA in August 1989, only one small insolvent thrift was placed with an acquirer, and none were liquidated. Doing nothing meant watching losses mount and watching asset values and franchise values deteriorate.

In the interim, also at President Bush's request, the Bank Board appointed conservators or receivers at all RAP-insolvent thrifts and brought in the FDIC as the managing agents for these thrifts.[25] By the passage of FIRREA, 262 thrifts were in the FDIC-RTC caseload. The justifications offered for this request were that FDIC management was necessary to bring these thrifts "under control" and that the FDIC (soon to become the RTC for thrift disposal purposes) would thereby become familiar with the insolvent thrifts so that the RTC could "hit the ground running" in disposing of the thrifts when the FIRREA legislation was finally passed. These thrifts, however, were already under tight supervisory control, and there was little the FDIC managing agents could do. In some instances, the FDIC's actions caused local interest rates on deposits to increase. As the data in Table 8-4 indicated, the Texas premium actually *increased* in March 1989.

With respect to the RTC's disposals of the insolvent thrifts, it is not apparent that the six-month period of FDIC management was especially beneficial in speeding RTC disposals. The pace, as of early 1990, had been disappointing. In its first seven months of operations, from August 9, 1989, through the end of March 1990, the RTC had disposed of only fifty-two thrifts with $17 billion in assets. The RTC did quicken the pace considerably in the second quarter, so that by the end of June 1990 the RTC had (since the previous August) disposed of 207 thrifts with $69 billion in assets. Still, its remaining caseload of insolvents on that date was 247 thrifts with $131 billion in assets.

The RTC's slow progress in disposals has been costly. It was the product of a number of factors: the RTC's unfamiliarity with thrifts, thrift assets, and thrift acquirers; the RTC's reluctance to provide acquirers with the types of assurances that the FSLIC provided in its 1988 transactions; the RTC's working capital difficulties; the FIRREA's restrictions on the flexibility of thrift operations, which reduced the franchise value of operating a thrift; the uncertainty by acquirers as to whether the Congress will impose further taxes and/or restrictions on healthy thrifts in the future; and the particular organizational structure that the FIRREA imposed on the RTC.

This last point is worth some elaboration. The FIRREA, in recognition of Congressional and Bush administration concern about the potential scope and power of the RTC's actions, created the RTC Oversight Board to set policy for the RTC and monitor and assess its actions. One consequence has been costly delay. Further, the RTC and the RTC Oversight Board are embedded in a larger structure of organizations and responsibilities created by the FIRREA. Figure 9-1 reproduces an organization chart developed by the RTC to represent those relationships. Chapter 3 mentioned that representations of the bank and thrift regulatory relationships sometimes look like Piet Mondrian's painting "Broadway Boogie Woogie." The FIRREA has clearly created another such representation for the thrift cleanup (though some cynics have suggested that Figure 9-1 more resembles an electrical wiring diagram).

The costly delays in disposals are likely to carry over to the RTC's efforts to manage and dispose of the large quantities of real estate and shoddy loans that it will acquire as it liquidates thrifts. The same relationship with the Oversight Board and with the other entities

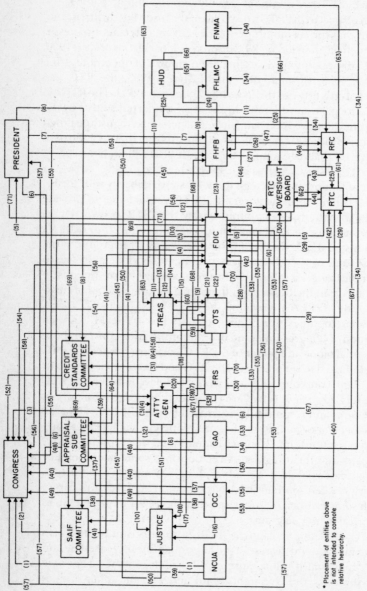

Figure 9-1. FIRREA: Functional Relationships

* Placement of entities above is not intended to connote relative hierarchy.

1. Enforce. Report § 918
2. Committee Report § 226
3. Enforce. Report § 918
4. Bridge Bank Activity § 214
5. Annual Reports § 501
6. Approve 2 Board Members § 501
7. Approve 4 Board Members § 702
8. Appoint 6 Comm. Members § 1205
9. Purchase FHLMC Oblig. § 731
10. Crim. Refer. § 918
11. Backup Funds § 511
12. Serve on BOD § 501
13. Preserve Minority Shops § 308
14. Report on Rec./Conserve. Activity § 212; Can Borrow 5B with Treasury's OK § 218; Quarterly Operating Plans & Forecast § 220
15. Supplement SAIF Funds § 211; FSLIC Fund § 215
16. Crim. Refer. § 918
17. Crim. Refer. § 918
18. Crim. Refer. § 918
19. Bridge Bank Activity § 214
20. Info on Holding Co. § 301
21. Approve Corp Debt Activity 222; Preserve Minority Shops § 308
22. Notify of Ins. Trans. § 206; § 301 New S & L Appl. § Assessments § 208; Enforce Action § 912; Subsidiary Activity § 220; Report on Rec./Conserv. Action § 212

23. Help fund SAIF § 211
24. Serve on BOD § 702
25. Reimburse RTC Property § 501
26. Help Fund § 511
27. Dist. Bank Allocation § 511
28. Serve on BOD § 203
29. Appoint Rec./Conserv. § 212
30. Serve on BOD § 501
31. Comm. Member § 1205
32. Comm. Member § 1101
33. Ann. Audit § 301 Audit of BIF, SAIF, FSLIC Funds; Old FSLIC Cases Ann. §219, 501.
34. Ann. Audit § 501, 511, 702, 731
35. Report on Rec./Conserv. Activity § 212
36. Civil Money Penalty Regs. § 907
37. Comm. Mem. § 1101
38. Comm. Mem. § 1205
39. Comm. Mem. § 1101, 1205
40. Reports and Appearances § 501
41. Report to BODs § 226
42. SAIF Funding Help § 211
43. Set Dir./Dispose Assets § 511
44. Reports/Funding Requests § 501
45. Rep. From Dist. Banks § 226
46. Send FSLIC Fund Money § 215
47. Issue Stocks § 511
48. Report Audit Findings
49. Enforce. Report § 918
50. Crim. Refer. § 918
51. Crim. Refer. § 918

52. Enforce. Report § 918; Bank Ser./Fee § 1002
53. Serve on BODs § 203
54. Ann. Reports on Fed. Financial Asst./Risk Assess. § 1403, 1404
55. Ann. Reports on Housing, Enforcement, Activities & Advances, Home Mortgage Disclosure Act § 721, 918, 1211 Title VII
56. Ann. Report on BIF, SAIF, FSLIC Funds § 220; Enforce. Report § 918
57. Ann. Report on RFC Activity § 511
58. Ann. Report on Activity § 301; Enforce. Report § 918
59. Oversee § 301
60. Preserve Minority Shops § 308
61. Issue Cert. § 501; Pay Under Special Cond. § 511
62. Oversee Activity § 501
63. Pay Back Assets on Dissolution § 511
64. Comm. Mem. § 1101; § 1205
65. Regulator § 731
66. Serve on BOD § 501
67. Bridge Bank Activity § 214
68. Request Advances § 714
69. Comm. Mem. § 1101; § 1205
70. Bank Holding Co. Approval § 208
71. Appoint 3 Board Members § 203

Source: RTC

shown in Figure 9-1 will surely add to costly delays. Further, the FIRREA specifically adds some extra restrictions on the RTC's actions: The RTC must give advance notice (at least ninety days) to nonprofit organizations, housing agencies, and lower income families of the opportunity to purchase certain types of residential properties. Finally, the RTC cannot sell any properties for less than 95 percent of "appraised market value" in areas of depressed real estate markets: Arkansas, California, Colorado, Louisiana, New Mexico, Oklahoma, and Texas.

This last restriction could become a serious impediment. Real estate appraisal is far from an exact science. If the appraisal on a property is at or below the true market price, this provision should not pose problems. But, if the appraisal is more than 5 percent above the true market price, or if the market falls between the time of the appraisal and the time of the expression of buyer interest, the RTC will not be able to sell the property. Delay, especially in the case of real estate, may well mean neglect of the property, physical deterioration, missed opportunities, and eventual sale only at a far lower price, which adds to the cost of the cleanup.

One cure for this problem is for the RTC's real estate managers to track the delays in the disposals of individual properties. Excessive delays in the disposals of properties[26]—excessive time spent in the RTC portfolio—should be seen as an indication that the original appraisal was too high and should automatically trigger downward-oriented reappraisals. Alternatively, as appeared to be the case in the spring of 1990, the RTC may be able simply to lower the benchmark percentage for properties that have been slow to move.

Finally, there is one easy measure for reducing costs that the RTC has neglected to take. As part of its 1988 disposals of insolvent thrifts, the FSLIC issued approximately $20 billion in ten-year notes to acquirers. Most of the notes carry interest rates that are linked to the interest costs of deposits in the state or region of the acquired thrift. The interest yield on the notes is exempt from taxation. Virtually all of the notes can be paid off (called in) by the RTC at any time; the FSLIC explicitly negotiated these call provisions into the contracts with acquirers to provide future flexibility for reducing the costs of these transactions. To pay off the notes, the RTC would have to borrow an additional $20 billion. But, by borrowing through the Treasury, the RTC would substantially reduce the interest costs that have to be paid on the $20 billion; and when the current note holders reinvest the $20 billion in assets with taxable income, the Treasury's tax collections will increase. Further, the FSLIC's notes were counted as part of the U.S. government's budget deficit and added to the national debt at the time they were issued in 1988. The substitution of U.S. Treasury debt for the FSLIC notes would thus have no budgetary implications. The RTC should pay off these notes as quickly as possible.

How Much Will it Cost?

The costs of cleaning up the problems of the insolvent thrifts will be horrendously large. The question of exactly how large those costs will be has two answers:

1. No one knows; only estimates can be made, which depend on future economic conditions and on future policy actions; and

2. The cost estimates depend on the measurement basis used.

Both answers are valid and warrant explanation.

1. *The role of future conditions and actions.* The costs of cleaning up the insolvent thrifts will depend on how many thrifts become insolvent and on how deep their insolvencies are—how large are the shortfalls between the values of their assets and the values of their liabilities. In turn, each of these considerations is related to the larger economic environment and to policy choices by the federal government.

The *number* of thrifts that become insolvent is partly related to their past actions and partly related to their current performance. To the extent that they made imprudent loans and investments in the past that the accountants have been slow to recognize, these embedded losses must eventually be recognized and the assets written down. These writedowns will decrease those thrifts' aggregate asset values and cause the net worths of many of them to become negative. Further, to the extent that real estate values in the Southwest and elsewhere continue to slide, these value reductions will also mean asset writedowns, reductions in net worth, and more insolvencies. At the same time, the economic environment (which will help shape the course of real estate values) will influence the direct operating results of thrifts. Any significant movements in the levels of interest rates will affect them since most thrifts still operate with longer average maturities for their assets than for their liabilities (though their maturity mismatch is far less than it was before 1980). The general level of competition among thrifts, and between thrifts and other providers of deposit liabilities and of mortgage loans, will also be an important determinant of the profitability of the thrift industry. The future actions of Congress, in continuing or modifying (in either direction) the FIRREA's taxation of and regulatory limitations on *solvent* thrifts will be a further influence on their profitability.

The *depth* of the insolvency of these insolvent thrifts—and hence the cost to the RTC of cleaning up each one—will largely be determined by the future course of real estate values. This is because the RTC's disposals have largely been liquidations, transfers of deposits, or placements that allowed the acquirers to take only a fraction of the assets of the insolvent thrifts. Thus, the majority of the assets of the insolvent thrifts are remaining with the RTC, to be sold over time. (If, by contrast, the RTC could dispose of all insolvent thrifts through placements with acquirers and if the acquirers would accept the assets at a specific up-front price, then the RTC's costs of that disposal would be known at the time of the transaction) Also, the RTC's disposals of all of its insolvent thrifts will extend over a few years, so actual and expected real estate values at those times will affect the RTC's costs of those disposals. Further, as was explained in Chapter 8, many of the FSLIC's 1988 transactions involved capital-loss coverage (and capital-gains sharing), so the final costs of those transactions will depend on the future course of real estate values.

Government policies, though, can also affect these costs. To the extent that the RTC is less effective or efficient in managing or selling individual assets, or in disposing of the thrifts themselves, the implied depth of the insolvency will be greater and the costs of the cleanup will be higher. Also, to the extent that acquirers find thrift franchises less attractive—either because of current taxes and restrictions or because of fears of further taxes and restrictions—the implied depth of the insolvency will again be greater and the RTC's costs will be higher.

In sum, the estimates of the costs depend on future economic conditions and policy actions. As those conditions and actions change, the estimates will change. And only after the last asset from the last insolvent thrift has been sold will the complete calculation of the costs be possible.

2. *The measurement basis.* Even if any statement of the costs is based on estimates, there are still three issues of measurement that require discussion. First, what starting date for the disposals should be used? Most of the estimates of the costs of cleaning up the

insolvents explicitly or implicitly start with the 1988 disposals and do not include any earlier disposals. As Table 8-2 indicated, the inclusion of the 1985–1987 disposals would add somewhat to any total.[27]

Second, most estimates do not include some or all of the costs to the Treasury of the forgone tax receipts that are likely to occur because of the tax breaks attached to the FSLIC's 1988 transactions. Again, as Table 8-2 indicated, these forgone tax receipts would add to the total.

Third, and most important, *virtually all of the estimates that have appeared in the media do not use a discount rate* to deflate future expenditures (e.g., payments to depositors or to acquirers) and receipts (e.g., sales of assets). Instead, they simply use the sum of all future cash flows, including the interest payments on borrowings, regardless of the timing of these flows. This appears to be the consequence of the Congress's budgeting procedures, whereby the Congress wants to keep track of aggregate receipts and expenditures and the years in which they occur, but is not sensitive to the time value of money within those aggregates.[28]

This use of undiscounted cash flow cost estimates is both confusing and distorting. It is confusing, because the nominal cost estimate will vary purely because of the time horizon chosen. Since the $50 billion in borrowing is for thirty-to-forty years and the interest payments on this borrowing is included in these cost estimates, a longer time horizon will yield a higher apparent cost. For example, at the time the FIRREA was passed, one set of estimates, which were based on a ten-year horizon, placed the costs at around $160 billion; another set of estimates at that same time, based on a thirty-year horizon, placed the costs at around $250 billion.

These estimates are distorting and misleading because they ignore the time value of money. Also, they double count by including the $50 billion in borrowing for expenditure on disposals (or, equivalently, the $50 billion to be repaid in thirty years) and then also including the annual interest on that borrowing (and neglecting to use a discount rate on any of these figures). This entire procedure ignores the very important idea that a dollar spent (or received), say, five years in the future is less costly (less valuable) than a dollar spent (received) currently. If a specific obligation could be satisfied by waiting five years and spending $10 then, rather than spending $9 currently, this delay would definitely be worthwhile at current interest rate (discount rate) levels; but the cash flow approach that ignores the time value of money would mistakenly conclude that delay was more costly.

The only sensible way to sum and compare dollars spent and received at different time periods is to use a discount rate to bring them all to the present. For the estimates of costs of cleaning up the insolvents, this method would involve the sums of the discounted expenditures to acquirers and depositors and the discounted receipts from the liquidations of assets. In essence, this "present discounted value" approach would yield an estimated sum of money that would need to be set aside at the present so that it, plus the interest on any unspent amounts at future points in time, would be adequate to cover all of the costs over time. This method would eliminate the problem of the varying time horizons, since it would (properly) ignore the interest payments on the borrowings as double counting. It would also indicate that any increases in nominal costs over time should be interpreted as real cost increases only if the nominal costs were growing at a rate that was higher than the discount rate.

This is the method that the FSLIC used in making its decisions with respect to individual disposals, and it is the method used to reach an estimate of $38 billion for the present discounted cost of the FSLIC's 205 disposals and 18 stabilizations for 1988.[29]

In early 1989 the Bank Board's field-force supervisors estimated that approximately another 400 thrifts would require disposal, at an estimated present discounted cost of about $40 billion. As of early 1989, then, the present discounted cost of cleaning up the insolvents— including the 223 disposed of or stabilized during 1988 and the 400 still to go—appeared to be about $80 billion.

The Bush administration, in developing the FIRREA borrowing proposals, requested $50 billion in new borrowing authority to clean up the remaining insolvents (rather than the $40 billion estimated by the Bank Board). In part, this $50 billion represented greater pessimism on the part of the Bush administration; also, since the $50 billion was a nominal amount that would not change over time, this larger amount was necessary because the RTC would only be able to borrow and spend it over a three-year period. During these three years the nominal amount of the costs of cleaning up the remaining insolvents—even if $40 billion was the correct amount in early 1989— would continue to grow as the insolvents' operating losses continued and would virtually exhaust the $50 billion by the time it was completely spent in 1992.

As of the summer of 1990, it appears that the estimated costs of the cleanup have probably increased to about $140–150 billion on a present discounted value basis. (The estimates using undiscounted cash flows and thirty-year or forty-year time horizons have increased to a range of $350–500 billion.) This increase appears to be due to at least five factors:

First, more than a year has passed since early 1989 and the initial development of the FIRREA legislation. Even if the costs that were estimated in early 1989 had not changed in nominal terms, the progression in time to early 1990 would raise the estimate in *present* discounted value terms by an amount equal to the discount rate.

Second, in the year that followed the proposal of the FIRREA legislation in early 1989 comparatively little cleanup activity occurred. In the interim, the insolvents continued to run large operating losses; to the extent that these operating losses mounted at a rate higher than the discount rate, this adds to the estimated costs.

Third, as Table 6-10 indicated, real estate values in the South and Southwest were still declining during 1989, which deepened the extent of the insolvencies (and would also add to the estimated costs of the FSLIC liquidations and placements with acquirers during 1988). Also, by 1990 the slump in real estate prices had spread to other regions of the United States.

Fourth, the operating environment for thrifts has been less favorable than had been expected, causing more of them to slide into insolvency or to be near-term candidates to do so. The continued decline in real estate values, the decline in the junk bond market (which the FIRREA contributed to, at least in part, by requiring that all thrifts must sell their junk bond holdings within five years[30]), the continued operation of hundreds of insolvent thrifts (which raises deposit costs), the higher taxes levied on thrifts by the FIRREA, and the regulatory restrictions imposed by the FIRREA have all contributed to this more difficult operating environment and the diminished economic prospects for hundreds of thrifts. As of the summer of 1990 the number of thrifts that were already insolvent or prime candidates to become so, and would thus require disposal at a cost to the RTC, appeared likely to be around 600–800, rather than the 400 that had been estimated in early 1989. (The costs of disposing of the additional 200–400 were unlikely to be as large as those involved with the initial 400, however, since the financial problems of the second group were likely to be less severe.)

Fifth, the FIRREA's taxes and restrictions on thrifts, as well as the prospects that the Congress may impose yet further taxes and restrictions in the future, have reduced the going-concern values of the insolvents to potential acquirers, thus increasing the RTC's costs of disposals.

In sum, in the environment created by the FIRREA, there should be little surprise that the costs of the cleanup have mounted. Depending on the future economic environment, the actions of the Congress, and the actions of the RTC, these cost estimates will likely change again in the future.

Who Should Pay?

In the public discussion surrounding the development and Congressional passage of the FIRREA, a popular refrain was, "Why should the public pay? Let the thrifts pay!" The Bush administration's formulation of the FIRREA, and the Act's final structure, represented a partial response to this outcry. The healthy remainder of the thrift industry was expected to bear approximately one quarter of the overall costs of cleanup; the Treasury (general tax revenues) would fund the balance.

Though the public's desire to have "the thrifts" pay the costs is understandable, it is a forlorn hope. At the beginning, it is important to remember that companies are only organizational forms that bring people together as customers, suppliers, employees, and owners. Thus, at best, it would be these people connected to thrifts who would bear the costs—in their roles as owners, through reduced profits; as employees, through lower wages and salaries; as suppliers, through reduced purchase prices for supplies; as depositors, through lower interest paid on deposits; and/or as loan customers, through higher interest rates and fees paid on borrowings.

Further, "the thrifts" can logically encompass only two groups of firms: the insolvents and the healthy remaining thrifts. As was discussed in Chapter 6, the moneys lost by the insolvent thrifts are not resting in safe deposit boxes in Switzerland; the losses arose through ill-conceived loans and investments and cannot be retrieved. Though the miscreant managers and owners of insolvent thrifts should be sued and indicted whenever legally appropriate, these actions will at best recover only a tiny fraction of the costs of the cleanup.

As for the remaining healthy thrifts, they are neither a logical nor financially feasible target to bear the costs. With few exceptions the remaining healthy thrifts did not engage in the high-risk activities of the insolvents; their greatest sin was the ill-conceived political lobbying that they and their trade associations conducted. Further, even if the American polity were to decide that *all* of the net worth of the healthy thrifts should be seized, the net proceeds would fall far short of the estimated costs of the cleanup. As of year-end 1989, the thrifts that were solvent on a tangible basis reported an aggregate tangible net worth of $44 billion. This was likely an overstatement of the market value net worths of these thrifts, and thus a seizure would yield even less in realizable value.[31]

In short, any hopes of extracting all, or even a major fraction, of the costs of the cleanup from "the thrifts" is illogical and unrealistic. And, as of early 1990, it appears that even the one-fourth burden envisioned by the FIRREA is unlikely to materialize. The industry's insurance premium contributions (based on the schedule shown in Table 9-1) were estimated on the expectation of a continuing 7.2 percent annual increase in thrifts' deposits (on which they would pay the premiums). A 7.2 percent annual rate of increase would appear to be reasonable for a stable thrift industry: If, say, the real growth in the U.S. economy is about 2 percent and the rate of inflation is about 5%, then the nominal economy would grow annually at a rate of about 7 percent, so the 7.2 percent expected growth in deposits would be in approximate consonance with the nominal growth of the economy. As of 1990, however, the overall thrift industry has not been stable. Deposits have been shrinking, not growing; between year-end 1988 and year-end 1989, thrift deposits *fell* by 2.7 percent. Until the RTC's inventory of current and likely insolvents are cleared away, significant growth in thrift deposits is unlikely.

As a consequence, the healthy thrifts' contribution to the costs of the cleanup are likely to fall short of the absolute sums that were expected. This shortfall, plus the larger costs that now seem likely, means that the industry's share of the burden will be well below one-fourth.

Why should the remaining burden devolve upon the Treasury, and thus on the general public? (Since the U.S. government is currently running large budgetary deficits, placing the burden on the Treasury really means borrowing extra sums and placing the burden on future generations.) As has been argued repeatedly in this book, the moneys are being used to satisfy the insurance obligations of the U.S. government to depositors. Since it was poor design and administration *by government* of that insurance arrangement that generated these costs, it is reasonable that the constituents of that government should bear them. Besides, there is no other logical place for the burden to rest. And it would be unthinkable for the U.S. government to renege on these obligations.

Notes

1. Passage probably would have taken considerably longer, but for one triggering event: The CEBA had imposed a one-year moratorium, which the Congress subsequently renewed for another year, on the ability of thrifts to leave the FSLIC and join the FDIC (and thereby benefit from the FDIC's substantially lower insurance premium levies). That moratorium was scheduled to expire on August 10, 1989. Since an important piece of the FIRREA was intended to continue confining the thrifts to a specific insurance fund and to tax them, this date became a crucial one for enactment.

2. The concern of the Bush administration and of many in Congress was that these borrowings not be "on budget" and thereby trip the Gramm-Rudman-Hollings budget deficit limitations. The compromise achieved the desired outcome. The first $20 billion was added to the (fast expiring) fiscal 1989 budget at a time when it was too late to trip Gramm-Rudman. The prospective REFCORP financing mechanism was "off budget" and therefore also outside the Gramm-Rudman limitations; see Note 16.

3. Starting in 1995, however, the FDIC can raise the premiums of either fund by 7.5 cents (per $100 of deposits) per year, to a maximum level of 32.5 cents, if it deems the reserves of the fund to be too low.

4. In essence, these moneys are the funds that were left over and not required to provide the defeasing of the principal amounts of the FICO bonds.

5. The field-force examiners and supervisors, who had been employees of the FHLBs, were brought on to the Treasury's payroll.

6. The appointed members are nominated by the President and confirmed by the Senate, for terms of three years.

7. Their terms are for seven years.

8. For direct equity investments, however, the FIRREA specifically limits state-chartered thrifts to the 3 percent of assets permitted to federally-chartered thrifts.

9. The press widely reported the efforts of Speaker of the House Jim Wright and of Senators Alan Cranston, Dennis DeConcini, John Glenn, John McCain, and Donald Riegle. Also, the reporting of the closeness

of Congressman Ferdinand St Germain to bank and thrift lobbyists was instrumental in his defeat for re-election in 1988. And Congressman Tony Coelho resigned after media reports surfaced of his financial dealings with a California thrift. See Jackson (1989), Pizzo, Fricker, and Muolo (1989), and Adams (1990).

10. See Note 3.

11. This argument does not negate the role that Congress must play in overseeing the expenditures. The Congress must be able to set and change broad policies with respect to that spending, in order to achieve the maximum cost-effectiveness of those expenditures; however, the Congress cannot treat the spending itself as discretionary—unless it is prepared to renege on the insurance contract.

12. Banks can either use the FDIC logo that they had previously used or use the new logo required for thrifts. It is likely that banks will stay with their former logo, in an effort to differentiate themselves and perhaps to perpetuate the notion that the thrifts insured by the SAIF are somehow less solidly backed.

13. The status of the deposits in federally insured credit unions is even less clear.

14. This now seems to be largely true; see Hirschhorn (1990).

15. See Note 12.

16. Ideally, these borrowings should have been done through the Treasury, so as to achieve the lowest costs, but they also should have been exempted from the Gramm-Rudman limitations. They should have been exempted for the following reason: The Gramm-Rudman ceilings are designed to place limitations on government spending in order to limit the government's ability to create new direct claims on resources (e.g., through defense expenditures) or its ability to redirect claims on resources (e.g., through transfer payments, such as Social Security). But, in the case of the payments to clean up the insolvent thrifts, the real resources have already been expended (in the construction of the ill-conceived office buildings, etc.). The borrowings are being used to honor the *existing* claims of the liability holders in those insolvent thrifts. Unless the federal government were prepared to see those claims extinguished—in essence, unless the government were prepared to renege on the insurance contract—the borrowing does not create any *new* claims on resources. (It does, of course, create an issue of who will pay the ultimate costs—who in the future will give up *their* claim to resources—in order to honor the insurance obligation. But that issue is different from the one of the creation of new claims.) See Woodward (1989).

17. In the RTC's fifty-two disposals through the end of March 1990 the acquirers accepted only one quarter of the assets of the insolvent thrifts; the remainder stayed in the hands of the RTC.

18. Again, the issue was whether the Gramm-Rudman limitations would be tripped.

19. The FIRREA did require that the Treasury conduct an eighteen-month study of deposit insurance reform, and risk-related premiums was one of the items that the Treasury was explicitly required to consider.

20. Arguably, after the FSLIC acquired warrants in some of its 1988 transactions, the Bank Board might face a conflict in developing or enforcing regulations because it might be tempted to skew the regulations in favor of thrifts in which it owned warrants. One solution is to obtain warrants in all thrifts; see Bernheim (1988). Further, since the deposit insurer is also a regulator, the apparent separation does not cure even this conflict.

21. The FDIC is the insurer for all commercial banks. The OCC is the primary regulator for national banks. The Federal Reserve is the primary federal regulator for state-chartered banks that are members of the Federal Reserve System; it is also the regulator of all bank holding companies. The FDIC is the primary federal regulator of state-chartered banks that are not members of the Federal Reserve System.

22. For example, the Bank Board, in its risk-based net worth proposal, developed a method for dealing with interest rate risk that the bank regulators had ignored; see also Benston (1984) and Fischel, Rosenfeld, and Stillman (1987).

23. Market value accounting is another topic that the FIRREA assigned to the Treasury for its eighteen month study.

24. *Lincoln Savings and Loan Association* v. *FHLBB*, 670 F. Supp. 449 (1987), 856 F. 2d 1558 (1988).

25. In a few instances, thrifts that were still RAP-solvent but that were sliding rapidly downhill were also included.

26. What is considered excessive should be geared to the type of property under consideration. The normal amount of time required to sell a single-family residence is considerably less than the time required to sell an office building or a resort hotel.

27. Disposals that occurred prior to 1985 were largely dealing with insolvencies that were caused by the interest rate mismatch problems of 1980–1982, rather than the credit quality problems in the years that followed.

28. It is also the consequence of the general lack of sensitivity to the time value of money shown by the Congress and by the media.

29. The FSLIC originally estimated its costs for the 205 disposals at about $31 billion; only some of the tax benefits were included in these calculations. The eighteen "stabilizations" were estimated to cost another $7 billion.

30. Some notable defaults, and the bankruptcy of Drexel Burnham Lambert, also contributed to the general decline in the values of junk bonds.

31. Taxation of thrifts' activities over time might conceivably yield significant sums if thrifts had "captive" customers (e.g., depositors or borrowers) on to whom they could pass these taxes. It does not appear, however, that either depositors or mortgage borrowers can be held captive by thrifts in significant quantities; the competition in these markets—the alternatives for depositors and borrowers—is too good, and is constantly getting better.

PART FOUR

Reforming Deposit Insurance and Bank and Thrift Regulation

Regulation, Deposit Insurance, and the Lessons and Analogies from Other Types of Insurance

The preceding chapters have laid out the story of the S&L debacle. There are important public policy lessons for the regulation of banks, thrifts, and credit unions that can, and must, be learned from this experience.[1]

This chapter begins by asking a fundamental question: Why should there be safety-and-soundness regulation of these depository institutions? The answer to this question links this regulation logically to implicit or explicit deposit insurance. The ways that formal insurance companies protect themselves are then outlined and the analogies between these protections and the practices (or their lack) in the world of deposit insurance and depository regulation are explored.

Chapter 11 will draw upon these analogies and on the experience of the thrift debacle to develop specific recommendations for bank and thrift regulation and for the operation of the deposit insurance system.[2]

Why Safety-and-Soundness Regulation and Deposit Insurance?

Why should depository institutions be subject to safety-and-soundness regulation?[3] Most companies in the U.S. economy are not subject to this form of regulation. What makes depositories special?

Depositories engage primarily in two activities: taking in deposits and extending credit.[4] In this latter area, banks, thrifts, and credit unions are not alone as providers of funds to the rest of the economy. Insurance companies, pension funds, mutual funds, industrial finance companies, factoring companies, mortgage conduits, and consumer finance companies are other financial intermediaries that extend credit; nonfinancial companies make loans to each other through trade credit arrangements; and households are direct lenders to the nonfinancial sectors. As Table 10-1 indicates, depository institutions in total accounted for less than half of the total credit extended and outstanding to the rest of the U.S. economy in 1984 and less than 40 percent in 1989; if equity investments as a substitute for debt had also been included, the overall share held by depositories

Table 10-1
The Sources of Credit Market and Trade Credit Debt Outstanding to the Domestic Nonfinancial Sectors of the U.S. Economy, 1984 and 1989

	1984		1989	
Lending Source	Amount (billions)	Share of Total	Amount (billions)	Share of Total
Commercial banks	$1,792	27.3%	$2,640	25.1%
FSLIC-insured thrifts	838	12.7	1,092	10.4
FDIC-insured thrifts	178	2.7	246	2.3
Credit unions	85	1.3	153	1.5
Total depositories	2,893	44.0	4,131	39.2
Insurance companies	738[a]	11.2	1,366[a]	13.0
Pension funds	502	7.6	810	7.7
Other financial sector lenders	592	9.0	1,415	13.4
Nonfinancial sector lenders	1,848[b]	28.1	2,812[b]	26.7
Total	$6,573	100.0%	$10,534	100.0%

[a] Includes $25 billion in trade credit in 1984 and $45 billion in 1989.

[b] Includes $596 billion in trade credit in 1984 and $699 billion in 1989.

Source: Federal Reserve.

would have been below one-third. Depository institutions are simply not special in this respect. And there is nothing inherent in the process of granting credit that would logically call for safety-and-soundness regulation. If lending decisions are imprudent, the owners and the creditors of the lending enterprise will suffer; but this set of consequences flowing from poor investment decisions holds true for any firm in a private enterprise economy. The lending function alone cannot justify safety-and-soundness regulation.

Depositories are special, though, in the type of liabilities that they issue. Their liabilities are primarily short-term deposits. The holders of these deposits treat them as money or near-money. The depositors expect the deposits to maintain their value (to be redeemable at par) and to be relatively liquid. As a consequence these deposits are a significant form of household wealth and of the means of payment by companies and individuals.

Until 1972, depositories were unique in offering this type of liability. Since then they have shared this position with money market mutual funds. But, as of year-end 1989, the liabilities of MMMFs were only slightly more than 10 percent of the combined deposits in banks, thrifts, and credit unions. Depositories, then, are special. Their liabilities make them different, and are treated by the public as different, from those of General Motors or of "Sam's Dry Cleaners."

Safety-and-soundness regulation is not normally considered to be necessary to protect the liability holders of nonfinancial companies.[5] But the special nature of deposits, along with their ubiquity in use and importance as household assets, provides an argument for safety-and-soundness regulation: as a means of protecting individuals and (small) businesses that are not financially sophisticated enough to protect themselves in their choice of depository institution. An equivalent way of stating this point would be that safety-and-soundness regulation is motivated by an *implicit* notion of deposit insurance or guarantees for deposits. A formal system of deposit insurance, then, replaces this implicit guarantee with an explicit one that provides some back-up assurance: Though a perfectly functioning safety-and-soundness regulatory system would avoid any losses to depositors, no system is ever perfect; however, with an explicit guarantee, even in the event of a regulatory failure, the depositors will remain whole. Alternatively, as has been a major theme of this book, one can reverse the order of the two phenomena and think of the insurance as providing the guarantee to the depositors and the safety-and-soundness regulation as providing the protection to the deposit insurer.

There is one further aspect of depositories that warrants them special treatment: While most of their liabilities are expected to be liquid, most of their assets are considerably less liquid. They are thus exposed to the problem of runs. Depositors who become nervous about the likelihood that their deposit claims will be honored — either because they doubt the institution's solvency (i.e., the excess of the value of its assets over its liabilities) or its liquidity (i.e., its ability quickly to satisfy their specific claims for cash) — are likely to rush to the bank to withdraw their money. Their fears that other depositors may arrive first, reducing their own chances of successfully withdrawing those funds, can hasten this process and lead to mass runs on an institution.[6] Since even a solvent bank will rarely have enough liquid assets on hand to satisfy the simultaneous demands of a significant fraction of its depositors, such runs will cause a crisis for the bank. Because its assets are illiquid, quick sale of those assets to obtain the cash to satisfy the depositors' demands for cash is likely to be difficult at best and to entail losses. A forced quick sale could well cause an otherwise solvent bank to become insolvent.[7] If it cannot obtain the necessary cash quickly to satisfy its depositors' demands, it will have failed to have honored its promise that those depositors' funds were liquid. The violation of that promise is a logical basis for closure of the institution, which would surely mean more delay in the sorting out of claims and in the depositors' receiving their moneys.[8]

The potential for runs on depositories, then, are a special problem for this type of institution. And runs could well become contagious. Suppose a run develops on bank A. The depositors of bank B may now become nervous. The run on bank A has made them aware of (or heightened their sensitivity to) the possibility of a run on their bank. They may have imperfect knowledge of bank B's finances or may simply fear that their fellow depositors may become nervous. A new run can thus develop.

Even a reasonably well-functioning safety-and-soundness regulatory system alone cannot deal with this problem. If depositors fear that *their* institution may be subject to a regulatory failure (albeit a rare, but not impossible circumstance), or simply that their institution may be subject to a pure fear-of-adequate-liquidity run by their fellow depositors, runs will develop. Again, they may become contagious.

An effective lender-of-last-resort, such as the Federal Reserve, that will lend quickly to a solvent depository institution experiencing a run is a solution to the problem for solvent but illiquid institutions. (The safety-and-soundness regulator will, presumably, deal properly with most potential insolvency problems.) But, if the lender-of-last-resort hesitates in lending, the problem is not solved. And if the lender-of-last resort must lend even when it is not sure of the solvency of the financial institution to which it is lending, then it has effectively become a deposit insurer. An explicit and credible deposit insurance arrangement can clearly have value in forestalling runs and/or supplementing the lender-of-last-resort.[9]

Depository institutions, then, are special; and safety-and-soundness regulation and deposit insurance are legitimate means of dealing with their special problems. It is noteworthy that government-sponsored deposit insurance has a history of more than 160 years in the United States: The first state-sponsored deposit insurance arrangement was attempted by New York in 1829.[10] Government regulation of banks was present even before that date. The governmental concern about the liabilities of depository institutions and about their stability is longstanding and reflects a genuine problem posed by these institutions.

Some Lessons from Insurance

Deposit insurance, whether it is explicit or implicit, can properly be considered to be a form of insurance.[11] It is best likened to medical or legal malpractice insurance or to third-party automobile liability insurance. The insurer is promising to third-party beneficiaries (depositors; medical patients; legal clients; motorists and pedestrians) that they will be made whole in the event that the parties carrying the insurance (depository institutions; doctors; lawyers; drivers) cause them harm (by not redeeming deposits; by failing to give them adequate medical care; by failing to give them adequate legal services; by injuring them in an accident). In essence, the insurer is covering the liabilities of the insured party. The party carrying the insurance is different from the beneficiary; and the former is the focus of the insurer's attention and efforts in trying to reduce its payouts. (This third party liability insurance contrasts, for example, with health insurance or homeowner's insurance, in which the insured party and the beneficiary are the same.)

Since deposit insurance is similar to other, familiar forms of insurance, it is worth considering the tools with which insurance companies protect themselves against risks.

1. *Collect information.* Insurance companies typically collect (and

share among themselves) large quantities of data from which they can make actuarial estimates of risk. They also try to collect information about their insureds in order to be able to estimate the risks of their specific insurance coverages.

2. *Education and information.* Insurance companies often offer information to their insureds, in order to encourage the latter to reduce risks (and thus reduce the former's costs). Life insurance companies take out advertisements that encourage their insureds to modify their behavior in ways (e.g., eat balanced diets, avoid drug abuse, get periodic medical checkups) that will prolong their lives (and delay the insurance companies' payouts). Auto insurance companies and their trade associations encourage their insureds to wear seatbelts, to avoid driving while intoxicated, and generally to drive safely. Fire insurance companies often work with industrial and commercial customers to reduce fire hazards.

3. *Establish rules of applicability.* A specific form of insurance coverage will have specific rules of applicability. For example, an insurance company providing fire insurance to a commercial customer may require sprinkler systems or fire extinguishers that meet certain specifications; a company providing marine insurance may insist on certain hull designs; auto insurance usually specifies whether the coverage is for driving that includes commuting or business use or whether driving only for leisure activities is covered. By establishing the rules and limiting the activities or situations that are covered, the insurance company limits its risks and potential losses.

4. *Establish deductibles.* Insurance arrangements almost always have deductible clauses, whereby the insured party is expected to absorb an initial portion of any loss. The deductible may be only one specific amount, or the insured party may have a choice of higher or lower deductibles that can be chosen in conjunction with the other terms of the insurance coverage.

Deductible clauses provide both direct and indirect protection for the insurer: The larger is the deductible clause, the larger is the direct absorption of loss by the insured party before the insurer must make payments to cover loss. Further, the larger is the deductible, the more careful the insured party is likely to be in trying to prevent losses.

5. *Co-insurance.* In addition to a deductible clause, some insurance arrangements may call for the insured party to cover a specified percentage of the losses beyond the deductible. As is true for the deductible, co-insurance provides both direct and indirect protection for the insurer: The greater is the percentage of co-insurance

absorbed by the insured party, the lower are the direct payments by the insurance company for any loss beyond the deductible. Also, the greater is the co-insurance percentage, the more likely the insured party is to exercise care.

6. *Charge premiums.* In the light of the expected risks, deductible, co-insurance arrangements, costs of administration, and competitive market conditions, insurance companies charge premiums for the insurance coverage.

7. *Re-insurance.* In instances of large or otherwise undiversified blocks of insurance coverage, insurance companies may choose to reduce their risks by selling or re-insuring a portion of the coverage with other companies.

8. *Limiting the amount of coverage.* Insurance companies often limit the amount of coverage that they will provide. Auto insurance liability policies, for example, usually have a maximum coverage for third party injury of $100,000 per person and $300,000 per occurrence (accident).

9. *Canceling or refusing coverage.* If an insurance company feels that its pricing and other measures do not protect it adequately, it can cancel or decline to renew its coverage or refuse to offer coverage to new applicants.

Exploring the Analogies from Insurance

Chapter 3 provided a summary of the structure of the safety-and-soundness regulation and deposit insurance rules that surround depository institutions. Since this safety-and-soundness regulation can be considered to be an integral part of the deposit insurance system, the list of insurance protections just presented can be used to explore the analogies and shortcomings that are present in the combined safety-and-soundness regulation and deposit insurance arrangements.

1. *Collect information.* Depository institution regulators regularly collect vast amounts of accounting information (balance sheet and income-expense data) from their insureds. These data are collected quarterly in great detail and monthly in somewhat sketchier detail. In addition, the regulators conduct regular examinations of their insureds. In principle, the regulators ought to be able to determine the risk profiles of their depositories from this wealth of information.

This information collection effort is fundamentally flawed, however, because it relies on an accounting model (GAAP) that is cost-based and backward-looking rather than on one that is focused on current market values.[12] The calculation of net worth is based on this accounting model. Net worth is the equivalent of a deductible; also, the regulators' abilities to restrict activities, to limit asset and liability growth (limit the amount of coverage), and ultimately to take control of a depository institution (cancel the insurance coverage) are geared to measured net worth. Thus, the regulators' abilities to gauge their exposure to risk and to take actions to reduce their risks are tied to this seriously flawed information system.

Further, the regulators have not systematically asked for "dynamic accounting" information. The risks to the insurance fund are related not only to the current (market value) net worth of a depository institution but also to the possible future values of the assets and liabilities that are already on the institution's balance sheet; assets (or liabilities) with the same current market value may have different values under varying future scenarios of interest rates and/or macroeconomic conditions. For example, a thirty-year fixed rate mortgage for $100,000 can be compared to a thirty-year adjustable rate mortgage for the same amount. If both mortgages are made at current interest rates, both will properly carry a current market value of $100,000. If interest rates rise in the future, however, the fixed rate mortgage will decline in value, while the ARM will decline less or not at all, depending on how closely its terms adjust to the change in interest rates. For the same reasons, interest rate changes will affect differently the future values of short-term (or adjustable) and long-term (fixed-rate) deposits or other liabilities. Similarly, defaults on loans to some types of commercial borrowers may be less sensitive to macroeconomic fluctuations than are loans to other types of borrowers.

By not obtaining this dynamic information, which would require simulations of the effects of these future changes in interest rates and macroeconomic conditions on institutions' balance sheets and income statements, the regulators are remaining ignorant of these latent risks.

Finally, surprisingly little use has been made in formal regulatory proceedings (rulemakings) of the data that have been collected.[13] For example, regulations concerning net worth standards or activities limitations have been based almost entirely on the horseback judgments of the regulatory agencies rather than on any systematic use of the wealth of data that is collected from their regulated institutions or that might be gathered through special studies or surveys.

2. *Education.* The regulatory agencies do make sporadic efforts to educate depository institutions' personnel on various banking issues. For example, the Federal Home Loan Bank Board made efforts in the early and mid-1980s (through the Federal Home Loan Bank System) to educate thrift managers about interest rate risk and about the new instruments that were becoming available to reduce thrifts' exposure to that form of risk; the Federal Reserve and its twelve individual banks periodically hold conferences, maintain research staffs, and disseminate that research.

The agencies, though, do not provide any systematic educational programs to their insureds, nor do the agencies have any formal educational requirements for the personnel of a depository institution (that would be comparable, say, to the requirements for obtaining a pilot's license).

3. *Establish rules.* Here is where the regulatory agencies excel, at least in quantity; the safety-and-soundness regulations that apply to depository institutions run to many hundreds of pages.[14] The regulations operate on the basis of a deeply flawed information system, however, and the agencies have made little use of available data as support for their rulemakings.

4. *Establish deductibles.* Since net worth has the characteristics of a deductible, the regulators' net worth standards for depository institutions should be considered to be the equivalent of a required deductible. But, again, the net worth standards are based on an inadequate accounting system.

The net worth standards that applied to depository institutions have traditionally been invariant to the perceived risks of the institutions. This uniformity changed in the 1980s and will change yet more in the 1990s. In 1984 the Bank Board modified its net worth standards to require additional net worth for a specific type of asset (direct equity investments) that it perceived to pose higher risks to the FSLIC insurance fund. In 1986 the Bank Board's revisions to its net worth standards included lower levels of required net worth for thrifts that met specific tests for reduced interest rate risk. The so-called Basel international capital standards, which will apply fully to commercial banks in 1993, are specifically risk-based in their orientation, since they specify different net worth levels for different classes of assets. The net worth standards that the Bank Board proposed for thrifts in December 1988 were similarly structured; they were modified and made final (in light of the mandates of the FIRREA) by the OTS in November 1989 and, like the commercial bank standards, will apply fully in 1993.

Unfortunately, the new bank and thrift risk-based standards continue to rely on the inadequate GAAP accounting system. Further, by setting net worth standards on individual asset categories, these systems focus only on the default risk and possible losses of assets in these individual categories.[15] The systems ignore the covariance effects (i.e., the value of some assets may tend to go up at the times that the values of other assets tend to go down, and vice-versa) that could cause the riskiness of a portfolio of assets to be less than the apparent riskiness of its individual asset classes.[16] Equally important, as of mid-1990 both systems ignore the interest rate risks that can arise because of maturity mismatches between assets and liabilities.

5. *Co-insurance.* Co-insurance has generally not been practiced with respect to the insured parties: the owners of depository institutions.[17] Co-insurance has sometimes been practiced, however, with respect to uninsured depositors (currently, the amounts above $100,000) and other creditors. When insolvent banks and thrifts have been disposed of through liquidations and depositor payouts (or transfers of deposits to other institutions), only insured amounts are covered; depositors with uninsured amounts and other creditors have shared proportionally with the deposit insurers in the insolvency shortfalls. Where the insolvent institutions have been placed with acquirers, though, all of the liabilities have usually been transferred to and have become the obligations of the acquirers.[18]

It is important to note that co-insurance with depositors is likely to create the problem of runs discussed earlier in this chapter. The greater is the degree of co-insurance, the more edgy depositors are likely to be and the higher the likelihood is that they will run at the first hint of unfavorable news about their bank or thrift.

6. *Insurance premiums.* The deposit insurance systems have always operated with statutory flat-rate premiums that have been insensitive to the risks of the insured depository institutions.[19] This legislative fiat has partially reflected the long-standing Congressional unwillingness to consider deposit insurance as an *insurance* system and has partially followed from the view that direct supervisory regulation could adequately control risks and that explicit pricing of risk through risk-sensitive premiums was neither necessary nor feasible. This latter view

is part of a broader political preference in the executive and legislative branches of the U.S. government for command-and-control regulation over pricing as a way to induce firms to modify their behavior.[20]

7. *Reinsurance.* The deposit insurers have never tried any reinsurance. Since the terms of the insurance arrangements have been specified by laws passed by Congress and by regulations authorized under those laws and promulgated by U.S. government agencies, it was logical that depositors would think of the U.S. Treasury as the de facto reinsurer. The FIRREA converted this de facto assumption into a de jure reality. But, despite the "full faith and credit" endorsement that the FIRREA gives to insured deposits (indirectly and imperfectly[21]), the Treasury's role is nevertheless limited. The Treasury is not yet an automatic and complete reinsurer for the deposit insurance funds.

8. *Limits on the amount of coverage.* The deposit insurance coverage that is applicable to the insured party—the depository institution— is roughly the aggregate amount of deposits in that institution. The nominal limit on the insured amount (currently $100,000) has had little relevance, for two reasons. First, household depositors have had the ability to establish separately insured deposits in the names of separate household members in the same institution[22] and also to establish separately insured deposits in other institutions. Also, pension funds, deposit brokers, and other fiduciaries can pool the funds of multiple beneficiaries, deposit them in a single ("jumbo") deposit, and retain coverage for the individual beneficiaries (so long as their individual amounts are below $100,000). Thus, it is not surprising that thrifts, which rely to a very large extent on household deposits, have had a very high ratio of insured deposits to total deposits. That ratio has never been below 90 percent[23]; in 1988 it was reported at approximately 92 percent.[24] Further, the amount of uninsured deposits in insolvent thrifts that were disposed of by the FSLIC were typically only 1-2 percent of deposits; and when placing insolvent thrifts with acquirers the FSLIC frequently transferred all liabilities (including uninsured deposits) to the acquirers.

Second, though commercial banks have been less dependent on household depositors and more dependent on commercial depositors, whose deposits (checking and large time deposits) are more likely to be above the insured amount (in 1988 the ratio of insured commercial bank deposits to total commercial bank deposits was approximately 75 percent), the FDIC has tended to treat all deposits as covered by insurance, especially in large banks that would be difficult to liquidate.[25] And, like the FSLIC, even in the case of placements of smaller banks, it has frequently transferred all liabilities to acquirers.

Thus, the limits on the insurance coverage provided to an insured party have been embodied in limits on the ability of a depository institution to expand its assets (and fund those assets with deposits). The net worth requirements have been the usual means of limiting the coverage provided by deposit insurers. As Chapter 7 indicated, however, in 1984 the Bank Board tried (unsuccessfully) to limit exposure by restricting the $100,000 coverage to any deposits provided by deposit brokers; from 1985 onward the Bank Board specifically limited growth by undercapitalized thrifts and by thrifts that were meeting the then-applicable net worth standards but were below the eventual ("fully phased-in") net worth goals.[26]

9. *Refusing or canceling coverage.* By refusing to grant a federal charter, or by refusing to grant insurance to a state-chartered depository institution, or by refusing to approve a change of ownership (change of control) to new parties, the federal regulators have the power to refuse new coverage; the standards for these refusals were discussed in Chapter 3. Once an institution has been granted coverage, however, cancellation has been considerably more difficult. The formal cancellation procedure has traditionally been a lengthy one and has been rarely used. (The FIRREA contains provisions to compress the necessary time considerably.) Instead, the appointment of a conservator or a receiver, by removing the owners and managers from control, has been the traditional means of canceling insurance coverage. Though other grounds for the appointment of a conservator or receiver exist, the insolvency of the institution is almost always the main justification used. But, because insolvency is measured according to GAAP (or RAP), the delay until this book insolvency has been reached has invariably meant losses for the deposit insurers.[27]

Additional Tools

Government regulators have an additional set of risk-reduction tools that do not have a ready analogy with those of normal insurers: measures that alter the gains and losses from risk-taking by depository institutions. The regulators can bring civil or criminal suits in response to perceived rules violations.[28] The FIRREA strengthened these powers and also gave the regulators the ability to levy civil money penalties for violations. The threat or actuality of these suits reduces the expected gain from risk taking and adds to the expected losses.

Further, in a number of the FSLIC's 1988 placements of insolvent thrifts with acquirers the FSLIC retained warrants equal (usually) to a 20 percent ownership position in the transferred thrift. Like a deductible clause or co-insurance, these warrants have both direct and indirect benefits for the insurer: The warrants will directly give the insurance fund a 20 percent share of any gains by these thrifts; and this 20 percent share for the insurance fund will reduce the thrift's expected gains from any risk-taking and thereby reduce somewhat the thrift's incentives to take risks.[29]

Notes

1. Most of the discussion in this chapter will focus on banks and thrifts, but the lessons apply to credit unions as well.
2. Both Chapters 10 and 11 draw heavily on White (1989a).
3. Safety-and-soundness regulation should be distinguished from consumer information and protection regulation (such as truth-in-lending information requirements) and economic regulation (such as limits on interstate branching); see Chapter 3.
4. Depositories, by providing checking accounts, also provide a payments mechanism. The trustworthiness of the payments mechanism, though, ultimately relies on the trustworthiness of the deposits.
5. The Securities and Exchange Commission's regulation of the financial markets consists primarily of information disclosure regulation, not safety-and-soundness regulation. Where other financial sector companies (e.g., insurance companies and pension funds) are subject to "prudential" regulation (the equivalent of safety-and-soundness regulation), the justification is the regulators' concerns about the *liability holders* of these companies: insurance claimants and pension beneficiaries. Thus, the justification for this regulation is similar to the rationale for safety-and-soundness regulation of depositories.
6. See Diamond and Dybvig (1983), Fischel, Rosenfeld, and Stillman (1987), and Gorton (1989).
7. See Kaufman (1988).
8. If banks could simply suspend their promise of payment until they could convert their assets into cash at a reasonable pace, this suspension would ease the bank's problems; however, it would mean an unilateral abrogation of the bank's "contract" with its depositors and would deprive them of the promised liquidity of their assets.
9. See Baer (1985), FDIC (1989), Carns (1989), and Murton (1989).

10. See Calomiris (1989a; 1989b).

11. The reasons for Kane's (1989, p. 5) claim to the contrary are unclear.

12. The RAP framework was generally a further departure from market values; see Chapter 5.

13. There is a sharp contrast between the general practice in the bank regulatory area and the practice, say, in the environmental regulatory area. The bank (and thrift) regulators typically provide little or no data support in justifying their regulations; by contrast, the Environmental Protection Agency (EPA) typically relies on extensive research studies and modeling efforts to justify its regulations. Though executive judgments are also ultimately an integral part of the EPA's regulatory decisions, the effort to develop the statistical support for these judgments is usually extensive; see White (1981).

   One of the ironies of the controversy surrounding Charles Keating and Lincoln Savings of Irvine, California, is that Lincoln's opposition to the Bank Board's limitations on direct equity investments generated the only instance in which the Bank Board provided extensive statistical support for its regulation; see Barth, Brumbaugh, Sauerhaft, and Wang (1989) and Benston (1989); see also *Lincoln Savings and Loan Association* v. *FHLBB*, 670 F. Supp. 449 (1987), 856 F. 2d 1558 (1988).

14. They can also be likened to the covenants in bond indentures or the restrictions in bank lending agreements.

15. Even within this system of categories, the agencies did not marshal any data to support or justify the specific risk weightings and capital requirements.

16. This reluctance to embrace a portfolio approach is reinforced by the GAAP accounting model, since GAAP is generally not designed to recognize the changes (up or down) in the market values of assets and liabilities.

17. Beginning in 1984, however, the FSLIC did require, in its approvals of the acquirers of insolvent thrifts and in the formation of thrift holding companies, that the owners agree to maintain the thrift's net worth at required levels; if enforced literally, this would have implied 100 percent coverage of losses by owners—100 percent co-insurance. Because many owners were unwilling to expose themselves to this unlimited and open-ended obligation, the FSLIC agreed, in some instances to cap the net worth maintenance requirement at a specific amount—the equivalent of an increased deductible. In 1987 the FSLIC in some instances replaced this requirement with a "pre–nuptial" agreement that specified that the FSLIC could take control of the thrift at a specified positive level of net worth (rather than having to wait for insolvency, or zero net worth). In 1988 the FSLIC formally replaced the requirement of the open-ended net worth maintenance requirement with the option of either the limited net worth maintenance obligation or the prenuptial

agreement. The prenuptial agreement would be especially useful for acquisitions by mutual organizations or by multiowner groups, where the ability to enforce any net worth maintenance or co-insurance arrangement would be difficult at best.

18. In states that have enacted "depositor preference" statutes, however, the insurers have often transferred only the depositor liabilities to the acquirers and exposed other creditors to losses. In the case of Continental Illinois in 1984, the FDIC injected substantial sums into the bank while protecting all creditors. The bank's management was changed, however, and the stockholders were severely diluted; see Sprague (1986).

19. In April 1988 the Bank Board proposed a mild form of risk-based premiums, which would have allowed better-capitalized thrifts to enjoy lower premiums. Because of uncertainties as to the Bank Board's legal authority to implement this proposal and because of a shortage of funds at the FSLIC, the proposal was never implemented.

20. This has been true for other areas of health-safety-environment regulation in the U.S. economy; see Mills and White (1978) and White (1981).

21. See Chapter 9.

22. Also, joint accounts and trust accounts receive separate insurance protection.

23. At least since 1939, which is the beginning year of the available data for this percentage.

24. The actual percentage is probably higher, since the multibeneficiary jumbo accounts of pension funds and deposit brokers are insured but are listed in the "above $100,000" category of uninsured deposits.

25. That has come to be described (improperly) as the "too big to fail" doctrine. Rather, it should be described as the "too big to liquidate for insurance purposes" doctrine, since the FDIC could (and sometimes does) technically close a bank (cause it to fail), wipe out its shareholders, and then place it with an acquirer, while preserving all creditors.

26. In the spring of 1990 the FDIC adopted similar growth limits for banks.

27. But see Note 17. In the case of Merabank of Phoenix, Arizona, which became insolvent in 1989, the OTS was able to obtain $400 million from its holding company, Pinnacle West, in return for releasing the holding company from its open-ended commitment (which, the OTS feared, might not be enforceable in court).

28. There is a weak analogy with insurance: Insurers can sue if they suspect civil fraud, or they can try to convince local prosecutors if they suspect criminal fraud.

29. The warrants can also be likened to either equity rights or convertible debt being given to debt holders.

CHAPTER ELEVEN

Fundamental Reforms for Bank and Thrift Regulation and for Deposit Insurance

"The cause of the problem is deposit insurance." This is an often-heard phrase that is offered as an explanation for the S&L debacle of the 1980s. It is at best a half-truth that fails to explain why the deposit insurance systems did not develop problems during their forty-five years of existence prior to 1980 or why the banking system in the 1980s avoided the excesses that engulfed the thrifts.

The cause of the problem was *not* deposit insurance—at least, not as a generic proposition. As Chapters 5 and 6 demonstrated, the cause of the problem was the creation in the early 1980s of a specific set of opportunities, capabilities, and incentives for risk-taking by thrifts that were reinforced by a set of perverse federal policy actions that weakened the safety-and-soundness regulatory system at a time when it needed to be substantially strengthened—and all of these events interacted, with an exquisite sense of timing, with movements in the price of oil and changes in the tax laws. In essence, the cause of the problem was the *specific way* in which the deposit insurance system for thrifts was structured and administered in the early 1980s.

Until the late 1970s the patterns of bank and thrift regulation, and the operation of their deposit insurance systems, could be fairly described in the following way: Banks and thrifts were constrained by their charters and their regulatory regimes to a relatively narrow set of lending and deposit activities and to locations of offices and branches within circumscribed geographic areas. These activities and locations were thought to be the special preserves of these institutions. They were buffered from competition from without by the special nature of their activities and from competition from within by limitations on the granting of charters, by geographic limitations on branching, and by Regulation Q's ceilings on the interest that could be paid on deposits. The narrow set of activities, the safety-and-soundness regulatory framework (enforced by systems of examiners and supervisors), the shelter from competition (which created an implicit franchise value asset and thus added implicitly to net worth[1]), a relatively buoyant and stable economy, and the general expectation that individuals of fiduciary integrity would operate these institutions were all parts of the formula for the stable banking systems and stable deposit insurance systems of that era.

This formula had dissolved for thrifts by the late 1970s, with the consequences that this book has documented. It may well be dissolving for banks in the late 1980s and early 1990s.[2] In any event, the economic world of the 1990s is fundamentally different from the pre-1980 environment for both banks and thrifts, and there is little chance that banks and thrifts can return to that stable, protected world of the 1950s and 1960s—nor can bank and thrift regulation. The formula of that era is no longer applicable. A new formula for depository regulation and deposit insurance must be found.

Unfortunately, the Financial Institutions Reform, Recovery, and Enforcement Act of 1989 does not provide the appropriate formula. The FIRREA affected commercial bank regulation and deposit insurance in only a minor way. More important, as Chapter 9 indicated, many of the FIRREA's changes to thrifts regulation represent punishments that unnecessarily restrict the flexibility of operations of healthy thrifts; and the FIRREA ignores many of the tools of insurance protection discussed in Chapter 10.

The major focus of this chapter will be a set of recommendations for the fundamental reform of bank and thrift regulation and the deposit insurance system that is concomitant with that regulation.[3] The goal of these reforms is to re-create a system of depository institutions that encourages them to be efficient and flexible providers of financial services to their customers and simultaneously places minimal risk on their deposit insurer.

These recommendations draw heavily on the analysis of insurance protection tools in Chapter 10. The recommendations are the product of the experience of the thrift debacle described in the earlier chapters of this book. But they are equally valid for bank regulation and are vital for avoiding future crises in the banking sector, as well as for thrift (and credit union) regulation, and also for avoiding the unnecessarily rigid and inefficient regulation that is a natural alternative for many who seek a safe and sound depository sector.

Many of these recommendations are major in scope and challenge the conventional wisdom held by most owners and managers of banks and thrifts and also by most legislators and bank and thrift regulators. They are likely to consider these proposals as unduly radical. These reforms are vital, however, for the construction of a new, viable formula for depository regulation and insurance in the 1990s and beyond. Indeed, these recommendations are actually quite conservative—conservative in comparison with the more radical proposals considered at the end of this chapter and conservative in the sense that they build and expand upon the existing structure of depository institutions, regulation, and insurance.[4]

Market Value Accounting

The accounting system that generates the basic financial information for bank and thrift regulation must be changed.[5] This is *the* most important reform. Generally accepted accounting principles (GAAP), the framework that is standard for banks and thrifts, is inherently flawed, because it is a historical, cost-based, backward-looking system of registering value, rather than a current-value-based system.

The structure and orientation of the information system is crucial because net worth (capital) is calculated from this financial information. To aid the following discussion, the basic balance sheet for a healthy bank or thrift (from Chapter 3) is reproduced here as Table 11-1. Net worth is the arithmetic difference between the registered value of the depository's assets and the registered value of its liabili-

Table 11-1
The Balance Sheet of a Healthy Bank or Thrift, as of
December 31, 199X

Assets	Liabilities
$100 (loans)	$92 (deposits, insured)
	$8 (net worth)

ties. Net worth acts like a deductible from the perspective of the deposit insurer: The larger is the arithmetic difference between the value of the assets and the value of the liabilities, the greater can be the fall in the value of the assets before the deposit insurer's obligation to the depositors is triggered. Many important regulatory approvals or limitations on activities and on expansion are keyed to net worth; the regulators' ability to take complete control of errant depository institutions has been largely limited to instances where the bank or thrift has already fallen into insolvency—zero or negative net worth.

This keying of regulatory behavior to net worth is quite sensible, *if* the calculation of net worth is an accurate representation of the deductible that is protecting the insurer. Because GAAP valuations are backward-looking, however, net worth calculations based on GAAP bear no necessary relationship to the *current* protection afforded the deposit insurer. Only a net worth calculation based on the current market values of the depository's assets and liabilities can measure that current protection.

Further, and even worse from the perspective of the deposit insurer, GAAP's cost-based orientation is likely to lead to a systematic bias toward a depository's reported GAAP asset values' being an *overstatement* of the market value of those assets. For many assets, if the asset increases in value (as compared with its GAAP book value), the institution can sell the asset in order to realize the gain (and invest the proceeds in new assets, which will have a higher cost basis on the institution's books). If the asset declines in value instead, the institution can continue to carry that asset on its books at its historical book value.[6]

In essence, the GAAP framework gives depository managers a tremendously valuable "option": They can sell their "winners" to show gains, while hiding their "losers" by continuing to carry these latter assets in their portfolio at original costs. Systematic behavior of this type[7], however, will mean that the assets that are on the books of the institution at any given time will tend to be the "losers" and hence their GAAP values will be overstatements of their market values; in turn, this will mean that their net worths are overstated. As markets for various types of assets of banks and thrifts become better developed (see Chapter 12), the ability of depository mangers to sell their "winners" to recognize gains will increase.

Instead of the existing GAAP framework, the bank and thrift regulators must move to a *market value accounting* framework, in which *all* of the assets, liabilities, and off-balance-sheet items of a depository institution are marked to market on a frequent basis. The net worth calculation that would be derived from these market value figures would thus be a much better estimate of the protection being afforded the deposit insurer. The frequency of this evaluation should be geared to the frequency with which the depositories report to the regulators. Quarterly market value reports would probably be adequate, though monthly would be better.

For some assets and liabilities (e.g., mortgage backed securities, stocks, bonds) that have functioning markets, price quotations from those markets could be used directly.[8] For others, such as home mortgages, commercial loans, consumer loans, and deposit liabilities, analogies with related markets where prices are quoted—mortgage-backed securities, bonds, and Treasury securities markets—could be used. For yet others, such as real estate holdings, periodic appraisals would be required.[9] The regulators would have to develop specific rules and formulas to guide these calculations.

In addition to these periodic "snapshots" of the current market values of the depository institutions' balance sheets, the regulators must also ask for "dynamic accounting" information: for market value simulations that would encompass the effects on the institutions' *existing* assets and liabilities of changes in the major economic influences (e.g., interest rates, macroeconomic conditions, and foreign currency exchange rates) on those institutions. Again, rules and formulas would have to be developed.

The arguments that are usually advanced against the collection and use of market value accounting information can be grouped into two categories[10]: (1) Either the information cannot be collected, because precise market values are not available and approximations would be required, or (2) it should not be done, because the information would reveal excessive short-term (transitory) fluctuations in balance sheet values (with implied fluctuations in reported income) that would somehow be misleading and would put depositories at a disadvantage vis-à-vis other firms in the capital markets.

At base, these arguments against market value accounting do not deal with the fundamental questions that an accounting information system must answer: What is the net worth calculation of Table 11-1 supposed to represent? If it is not designed to represent the market values that are the protection for the deposit insurer, why should the regulators (or anyone else) pay any attention to it?

Further, none of these arguments can survive close scrutiny. The apparent precision of GAAP is a snare and a delusion. At best, it is a precise tracing of historical cash flows.[11] This tracing may satisfy a stewardship notion of accounting, but it is only an *approximation*—and an increasingly poor one, at that—to the concept that matters: market value. A market value accounting framework, though it too would involve approximations, would be aiming at the correct target; increased use and practice would surely help reduce or eliminate anomalies.

Though a system of market value accounting would initially be more costly than the existing system, repeated use would lower these costs considerably, and banks and thrifts would tend to alter their portfolios so as to reduce those costs yet further. Securities firms and pension funds have learned to live with market value accounting; banks and thrifts would too. Further, market value accounting currently occurs every time that one depository buys another, and it is practiced by many depository institutions for their own internal managerial uses (but they do not want to reveal that information to their regulators).

It is true that market value accounting might initially reveal fluctuations in portfolio values; after all, bank and thrift managers have built their asset and liabilities structures for a world of cost-based GAAP accounting. But over time, managers would restructure those portfolios and adopt hedging and smoothing strategies; in the interim, the regulators need to know about those market value fluctuations (not all of which may be cyclical).

It seems highly unlikely that market value information would place depository institutions at a disadvantage vis-à-vis other firms in the capital markets, especially after the transition to less volatile portfolios had been made. Market analysts are constantly trying to extract market value information from the existing GAAP information that banks and thrifts report publicly. Analysts would likely welcome and reward firms that made this information more transparent, especially if there were a standardized market value accounting framework for all depositories on which analysts could rely.

The arguments against market value accounting, at best, are arguments for a transition period (at most, a few years), during which time banks and thrifts would report the market value information only to the regulators[12]; after the transition period it would be reported to the public as well.

There is a second reason to favor market value accounting, rooted more in political economy.[13] It is clear that the deposit insurers and regulators have been too slow to close insolvent institutions.[14] Partly, the absence of market value information has masked true insolvencies from the regulators. As Chapter 5 indicated, however, the Congress and the regulators were instrumental in muddying to a further degree what was already an inadequate accounting system for thrifts. Also, political pleas for forbearance for favored depository institutions have been, and surely will continue to be, made.[15] The consequence of these delays in closure have been additional costs in disposing of these insolvent institutions.

By making insolvencies more transparent, market value accounting would make politicians' and regulators' efforts to delay closure more difficult, or at least require them to offer better explanations to justify their actions.

The institution of market value accounting would mean a major change in focus and ethos for the accounting profession, as well as for the bank and thrift industries. All parties would have to re-orient themselves away from the tracing of historical costs and toward the measurement of current market value. This re-orientation will not be easy, and there will be costs. But the benefits surely exceed the costs. It is a vital reform.

Net Worth (Capital) Requirements

The actions of the bank and thrift regulators in the late 1980s and early 1990s to set higher and risk-based net worth standards for banks and thrifts were clearly in the right direction. But much more needs to be done.

First, and most important, the net worth standards must be based on a market value accounting framework. Any other accounting framework yields a net worth calculation that is a misleading representation of the protection that is available to the deposit insurer.

Second, the risk-based aspect of the net worth standards needs to be improved. Primarily, risk should be measured on a portfolio basis, rather than on a category-by-category default basis. Further, interest rate risk must be explicitly incorporated into the net worth requirements, even if the asset-by-asset default risk approach is retained. This is vital for thrifts[16] and will become increasingly important for banks since the risk-based net worth standards for banks will provide them with a strong incentive to increase their holdings of mortgages and mortgage-backed securities.[17]

Third, the regulators must resist any potential pressures to modify net worth requirements to satisfy other social goals; they must resist pleas such as, "The construction of more multi-family housing is a worthy social goal, so multifamily housing loans should have a lower net worth requirement." The net worth requirements should not be an instrument for advancing other social policies, but rather should be based solely on the risks to the deposit insurance fund; other social goals are best handled directly by the Congress through explicit programs.

Fourth, the regulatory agencies must undertake the necessary statistical studies to demonstrate the risks to which banks and thrifts are exposed and the appropriate net worth requirements that would follow. These studies are necessary to justify the levels and gradations that would apply either to the portfolio approach to risk or the asset-by-asset approach. The current standards have no such support.

Fifth, the required net worth level should be treated as the "bare-bones" minimum deductible that will usually, but not always, protect the deposit insurer from loss. Banks and thrifts should be encouraged to operate with even *higher* levels of net worth, which would further *reduce* the insurer's probability of loss for any level of risk embedded in a bank or thrift's portfolio. An insurance premium schedule that promises lower premiums when net worth levels are higher would provide the right type of incentive.

The opposition to risk-based net worth requirements has come from those who believe that the regulators will incompetently miscalculate risks and thereby require inappropriate net worth levels and/or will use the gradations in net worth levels to favor specific types of "socially desirable" activities irrespective of true risk.[18] Their preferred alternative is to have uniform net worth levels at relatively high levels (buttressed with market value accounting and easier cancellation of insurance).

This is a curious criticism. Uniform net worth requirements (as a percentage of assets) are not neutral; in the presence of differential risks, uniform requirements create their own distortions. And the requirement of higher net worth levels is not costless, since equity or subordinated debt (which would count as net worth for these purposes because the interests of these debtholders would be subordinated to the interests of the deposit insurer) are more costly means of raising funds than are insured deposits. It is more sensible to encourage the regulators to try to measure risk in the best way possible, and to set net worth standards accordingly, rather than to ignore it.

Risk-Based Insurance Premiums

The existing flat rate insurance premiums should be scrapped and replaced with a system of risk-based premiums. The Congress and the Bush administration must stop thinking of the premiums as a revenue raising mechanism and think of them instead as *insurance* premiums that should be set on an actuarial basis. The risk elements that influence the level of premiums should be the same as those that guide the required level of net worth, *plus* an element that is based on the actual level of net worth (determined according to market value accounting) of the bank or thrift. Since net worth is the equivalent of a deductible that protects the deposit insurance fund, the insurance premium schedule should reward and encourage banks and thrifts that choose to operate with net worth levels that are above the minimums.[19] A risk-based premium structure is thus not redundant in the presence of risk-based net worth standards; the two are complementary.

The principle that insurance premiums are lower when the insured party poses less risk to the insurer and/or takes out a larger deductible is one that is practiced in virtually all other forms of insurance. Its application to deposit insurance is long overdue.

The base on which insurance premiums are levied, even if they remain flat rate, should be broadened to include other liabilities (besides domestic deposits) of banks and thrifts where the liability holders (lenders to the bank or thrift) have a prior claim on the assets of the institutions.[20] These liabilities include the collateralized borrowings of banks and thrifts, such as mortgage-backed bonds and reverse repurchase agreements. Where the borrowings are collateralized (and periodically marked to market), the lender is virtually guaranteed to receive 100 cents on the dollar, while the insurer absorbs all of the losses from the institution's insolvency. The same is true of the recourse obligations on assets that are sold by a bank or thrift with recourse by the buyer in the event of loss. In these instances, the deposit insurer is effectively insuring these liabilities of the institution—but is not currently receiving any premiums on these liabilities. Equivalently, the insurer is giving up priority in its claims on the assets of the bank or thrift, but not receiving any compensation in return.

These liabilities impose risk on the insurer. The insurance premium base should include them.[21]

A move to an actuarially based insurance premium structure, which would focus on prospective risk rather than on revenue gathering, should be an important aid to banks and thrifts that want or need to raise fresh capital that would add to their net worths. A risk-based premium structure would reassure the capital markets that well-run banks and thrifts are less likely to have their insurance premiums raised and thus reassure investors that their investments are less likely to be arbitrarily taxed away in the future. In the case of thrifts, this reassurance would be enhanced by an indication from Capitol Hill that the Congress does not intend to levy new taxes on investors in other ways, such as taxing further the thrifts' equity positions in the Federal Home Loan Banks.

The opposition to risk-based insurance premiums comes from two camps. "Practical" members of the Washington policy community profess puzzlement over how to measure the risks that would enter a risk-based premium schedule. Why they have readily embraced risk-based net worth standards but are reluctant to base premiums on the same risk elements remains a mystery; their reluctance to gear premiums to net worth levels is also a mystery. The academic opposition to risk-based premiums is largely based on the same arguments as the opposition to risk-based net worth requirements,[22] and the same responses are appropriate.[23]

Prompt Disposal and Early Intervention

The bank and thrift regulators must take effective control of depository institutions when they reach insolvency, or preferably even before that point, and rapidly dispose of them. Delay is far too costly to the insurance fund.

The appointment of a receiver is the most powerful tool in the regulators' arsenal. It removes the owners from control and eliminates their ownership interests.[24] It is the effective or de facto removal of the third-party (deposit liabilities) insurance coverage of the owners; equivalently, it is the "failure" of the institution.

Though there are other criteria for the appointment of a receiver, the regulators have almost always relied on the insolvency of a bank or thrift as their primary criterion; however, the appointment of a receiver at the time of insolvency is not mandatory. As Chapters 5–7 documented, the Federal Home Loan Bank Board allowed hundreds of insolvent thrifts to remain in operation throughout the 1980s. The absence of any net worth meant that these thrifts had strong incentives to take risks. Though direct supervisory controls could and did limit risk-taking at most of these institutions, especially after 1985, these controls were not perfect. And, even with tight supervisory controls on prospective risk-taking, the managers of insolvent institutions had few incentives to prevent further deterioration and erosion of existing asset values.

The lesson of this experience is that the onset of insolvency (or preferably earlier) must trigger the institution's failure and the simultaneous appointment of a receiver (which removes the owners' interests) and disposal of the institution, either through liquidation or placement with an acquirer. In essence, at insolvency the institution must be closed—either permanently, as in a liquidation, or briefly (e.g., over a weekend) and then reopened under the new auspices of an acquirer. Delay in disposal means, at worst, continued risk-taking by the insolvent institution's owners and managers (if supervisory controls are absent or less than completely effective) or, at best, a steady deterioration and erosion of values by a "caretaker" management. Either route is costly to the insurer.

No bank or thrift should be "too big to fail" under this approach. Some may be "too big to liquidate"; but they cannot be too big to dispose of through placement with an acquirer.[25]

The principle of eliminating ownership interests[26] and removing senior managers (who were responsible for the insolvency) through these rapid disposals is an important one to maintain. Owners and managers must understand that they lose *all* when their bank or thrift has to be disposed of and the insurer's money has to be used to honor the obligations to depositors. If the regulators were to do otherwise, then all discipline and disincentives to risk-taking would vanish.

The owners and managers of insolvent or soon-to-be insolvent institutions often raise the possibility of "open bank" or "open thrift" assistance by the insurance funds. If such requests imply the injection of federal funds while the previous owners retain some or all of their ownership interests and the managers retain their positions, these requests should be rejected summarily. Such actions truly would be "bailouts" and would only encourage more risk-taking and insolvencies in the future. If, instead, they are pleas only for easing the process of placing these institutions with acquirers and ensuring the continuity of operations, then they should probably be treated more sympathetically.[27] But the principles of wiping out the previous owners' interests and removal of managers is an extremely important one to maintain.

Unfortunately, even the prompt use of the insolvency criterion for appointing a receiver usually means that the regulators have waited too long and have added to the costs of the insurer; this is especially true when insolvency is measured according to cost-based accounting systems such as GAAP or RAP. By the time an institution has reached insolvency on these historically oriented accounting bases, its market value is usually substantially negative, and the deposit insurer will experience a substantial loss.

The FIRREA makes the formal procedures of removing insurance coverage from a bank or thrift (which is entirely different from the process of appointing a receiver) easier and more rapid. But this formal procedure is relatively unfamiliar and untried. The FIRREA also added new grounds for the appointment of a conservator or receiver, which include prospective losses by a bank or thrift or other actions that could lead to insolvency. This is a move in the right direction, but more should be done. A legal strengthening, through regulation or legislation, of the other primary grounds for appointing a receiver—operating in an unsafe or unsound condition, or the dissipation of assets—would allow regulators to intervene earlier and thereby reduce the likely losses to the insurer. This early intervention would even have value in a market value accounting framework because the reported market value information will never be perfect and is likely to lag true market values when they are moving downward.

The "prenuptial" clauses that the Bank Board negotiated into many of its 1987 and 1988 transactions and holding company approvals provide an example of one possible way to proceed. These clauses gave the FSLIC (or its successor) the ability to take control of a thrift at a specified *positive* level of net worth. In these instances, the insurer does not have to wait for the thrift's insolvency. Following this example, the regulators might promulgate regulations that would indicate that a decline in a bank or thrift's net worth to a specific level would be considered a dissipation of assets and/or an indication of operating in an unsafe or unsound condition.[28]

The opposition to early intervention focuses on the argument that such action constitutes a "taking" of property from the owners of banks and thrifts. From an economics perspective, this claim is unlikely to have much validity. First, any early intervention process should allow the owners and managers an opportunity to increase the net worth of their institution (and thereby take it out of the "danger" zone) by raising fresh capital; this is the way that the FSLIC's prenuptial clauses were structured. If the owners and managers cannot convince the capital markets that their enterprise is a viable one, the owners' claim as to any significant valuable property is open to serious question. Second, a bank or thrift's GAAP net worth frequently overstates its market value net worth, especially when it is sinking toward insolvency. The FSLIC's historical experience certainly reinforces this point. The owners of a bank or thrift with no market value net worth would only be losing the value of the regulators' forbearance.[29] Finally, in the rare event that positive market value might be present when the insurer disposes of an insolvent institution, that value could (and should) be refunded to the owners.[30]

The Extent of Deposit Insurance Coverage

The coverage of deposit insurance should be extended to include *all* deposit obligations of insured depositories.[31] This 100 percent coverage would be the only way to deal conclusively with depositor runs. The continuation of partial coverage is a continued invitation to runs by nervous depositors, and proposals to cut back the extent of deposit insurance would exacerbate this problem.

Complete coverage of deposits does not mean, in any sense, an endorsement of a "too big to fail" doctrine. As the discussion of prompt disposal and early intervention indicated, prompt disposal is a necessity for all insolvent institutions, and size should not be a criterion.

This proposal for 100 percent coverage runs counter to many of the current suggestions concerning deposit insurance reform, which advocate reduced coverage. Some would simply roll back coverage per insured deposit to, say, the pre-1980 level of $40,000. Others would limit coverage to $100,000 (or a lesser amount) per individual for the aggregate of all deposits owned by that individual in all insured depository institutions. Yet others would provide coverage for some basic amount and then require that depositors absorb some portion (e.g., 20 percent) of any amounts above the minimum (i.e., they would be asked to co-insure on these larger amounts).

The idea behind the limitation of coverage is that depositors would become more cautious in providing the funding for high-risk ventures by the owners and managers of banks and thrifts. This wariness about possible loss would bring "market discipline" to the funding of these banks and thrifts, as it does to other firms that borrow money.

This approach ignores the special nature of deposits, who holds them, and why, which was addressed in Chapter 10. At best, this market discipline approach might make potential depositors more wary of *transferring* their money to depositories that are trying to expand rapidly. But depositors generally are not well equipped to monitor the financial actions and health of depository institutions. Even the regulators, who are supposed to have far more expertise in this area, have had an extraordinarily difficult time doing so. To the extent that their deposits are uninsured, depositors' fears of losses on *existing* deposits in institutions will surely heighten their sensitivity and nervousness and greatly increase the likelihood of runs. The more drastic is the proposal for reduction in deposit insurance coverage, the more serious are the consequent runs possibilities.

Further, the individual proposals for reduced coverage have extensive operational problems, aside from their likelihood of exacerbating depositor runs. A simple reversion to, say, $40,000 per insured deposit with or without co-insurance will mainly cause more transactions costs, as many depositors who currently invest in $100,000 certificates of deposits in a single bank or thrift will instead break that amount into three parts in three depositories. The limitation of $100,000 (or a smaller amount) per individual would still allow multimember families to enjoy a multiple of that amount, and clever depositors would surely find other ways (e.g., partnerships, dummy corporations) to obtain multiple coverage; enforcement could well be a nightmare.[32]

In sum, the proposals to reduce coverage would either lead to more transactions costs or prove unenforceable; and to the extent that they do succeed in reducing the insurance coverage on some deposits (e.g., on corporate checking accounts and short-term jumbo CDs), they will exacerbate the problem of depositor runs.

There is one other benefit that would flow from 100 percent coverage. The tradition of secrecy with respect to examination reports, investigations, disciplinary actions, and other regulatory actions vis-à-vis individual depositories has a long history. That tradition has been based largely on fears that revelation of this information would spark massive depositor withdrawals. The tradition of secrecy did not disappear in the 1930s, with the onset of deposit insurance, because runs were (and are) still a possibility. With 100 percent deposit insurance, the fear of runs could no longer be used to justify secrecy, and many more regulatory actions could become transparent and open to public scrutiny.[33]

Expanded Use of Long-Term Subordinated Debt

A way that market discipline could be brought to bear on banks and thrifts, without generating runs, would be through the expanded use of long-term subordinated debt: bonds that are issued by the depository and are explicitly subordinated to the interests of the insurer.[34] In the event of the insolvency of the bank or thrift, the subordinated debt holders would bear losses (after the equity holders' stake had been wiped out) before the insurer would have to make good on its obligations to depositors.

The advantages of long-term subordinated debt would be as follows: The interests of the debt holders in curbing risk-taking by the managers and owners would be similar (though not identical[35]) to those of the deposit insurer; the debt holders would not get any of the upside benefits of risk-taking, but would bear the down-side consequences (up to the amount of their investment) of that risk-taking. They would surely insist on covenants in their debt instruments that would give them powers vis-à-vis the institution's management to restrict risk-taking. The existence of this debt, then, would create an additional monitoring and enforcement group, as a supplement to the regulators; but, because it would be long-term debt, the debt holders could not run on the bank or thrift.

Another way of expressing the idea underlying the use of subordinated debt is that these debt holders are a group of private co-insurers who bear the first losses after the equity owners. Suppose that a depository had issued long-term subordinated debt equal to 4 percent of its assets. From the perspective of the government deposit insurer, the effect would be similar to the institution's arranging for private insurance to cover the first 4 percent of losses to depositors. The subordinated debt method, however, has some advantages relative to private insurance. First, all of the private "insurance" money is made available up front; the government deposit insurer does not have to worry about the private insurer's ability to cover any loss. Second, because the debt is long term, the private "insurers" cannot run on the institution. Third, the legal rights and privileges of subordinated debt holders, and their ability to develop covenants vis-à-vis an institution's management, are relatively well developed[36]; the legal rights and privileges of a private deposit insurer that operated alongside a government insurer would have to be developed from scratch.

The interests of the insurer would be best served by having this debt in the hands of knowledgeable institutional holders, such as pension funds, insurance funds, and mutual funds. Since it is a sophisticated instrument, it probably should not be sold in small denominations to individuals over the counter in the depository.[37] Also, to the extent that this debt is publicly traded, the insurer could use the price and yield information to learn about the market's assessment of the prospects of the institution. The pricing of this debt, though, would partly depend on how promptly the regulators exercise their call option to appoint a receiver and dispose of an insolvent institution.

Because subordinated debt is a protection for the deposit insurer, it should be counted as net worth for a bank or thrift, so long as it is of long-term maturity;[38] because it brings a supplemental group of beneficial monitors and enforcers, its use should be encouraged—or perhaps even required. Though a net worth that consisted almost entirely of subordinated debt would not be desirable,[39] a net worth structure that was composed of, say, up to 50 percent of long-term subordinated debt would be quite reasonable. Table 11-2 provides a sample balance sheet for a healthy depository that would have 25 percent of its net worth structure as subordinated debt.

Unfortunately, the sentiment among bank regulators is distinctly hostile to the use of subordinated debt. They see this instrument primarily as *debt*, which requires interest payments and the eventual

Table 11-2
The Balance Sheet of a Healthy Bank or Thrift with
Subordinated Debt, as of December 31, 199X

Assets	Liabilities
$100 (loans)	$92 (deposits, insured)
	$2 (subordinated debt, net worth)
	$6 (equity, net worth)

repayment of principal. Their strong preference is for owners' equity as net worth.

This hostility is unfortunate. It is true that equity is a more flexible "cushion," since dividends can be more readily suspended and there is no principal to be repaid. But equity investments do not come to banks and thrifts for free. Investors will provide equity capital to these institutions only if they expect an adequate return on their investment. That expected return will be higher on equity investments than on subordinated debt investments, because the former is more risky. The long-run drain on an institution's earnings must be greater for equity investments than for subordinated debt. Further, equity holders are the ones who benefit from the upside of the risk-taking by their management; the absence of subordinated debt holders means the absence of a restraining influence on management.

The choice of equity versus subordinated debt as protection for the insurance fund, then, involves a tradeoff: greater flexibility in cushioning versus lower long-run costs and an extra restraining influence on managements. The optimal terms of the tradeoff are unlikely to be satisfied at the outcome of 100 percent equity and 0 percent subordinated debt. A mixture of the two should be encouraged.

The Treasury as Backup

The U.S. Treasury must become the automatic and unquestioned financial backup for the deposit insurance funds. The Congress needs to be educated to the full implications of the operations of federal deposit insurance. As has been stressed throughout this book, the recent expenditures of the insurance funds in relation to insolvent institutions are not properly described as "bailouts," but are instead the honoring of the insurer's obligations to depositors—either directly through payouts to the depositors or indirectly by paying an acquirer to assume the obligations to depositors. Once the Congress has put "the full faith and credit of the United States" behind insured deposits, those payments are not discretionary, and the Congress cannot continue to think of them as somehow subject to the whims of the budgetary appropriations process. The Congress must, however, monitor and oversee these expenditures, in order to ensure that the funds are spent in a cost-effective manner.

The fears of insured depositors in insolvent thrifts in 1987–1989 that they might not receive speedy and full payment caused deposit interest rates to be higher than they otherwise would have been, which raised the FSLIC's and RTC's eventual costs (and thus the U.S. Treasury's costs) of disposing of these insolvents (as well as raising deposit costs for healthy thrifts). These costs could have been avoided, and could be avoided in any future insolvencies, if the Congress had clearly designated the Treasury as the automatic backup for the insurance funds.

Further, speedy disposal of insolvents is vital for reducing the costs of the deposit insurer. With an automatic Treasury backup in place, the regulators should never again be forced to "warehouse" hundreds of insolvent institutions because of inadequate moneys in the deposit insurance fund.

The Congress must come to understand that the claims of insured depositors in insolvent banks and thrifts are entitlements and must be treated in the same way as are payments of interest on the national debt or payments to Social Security claimants. These latter types of payments are not treated as discretionary; the Congress has passed a permanent authorization for the Treasury to make these payments as is necessary. The same must be done as a backup for the deposit insurance funds.

Allowable Activities: The Depository, the Holding Company, and their Financial Relationships

With better information, improved net worth standards, risk-based premiums, and improved intervention powers in place, the Congress and the regulators should reassess what activities (e.g., types of lending or other services) ought to be permitted to take place within the domain of the insured entity and what ought to be relegated to the holding company.

The basic criterion for the inclusion of an activity within the domain of the insured institution should be the capability of the field force of examiners and supervisors to assess its riskiness (preferably, in a portfolio approach) and to be able to monitor its performance and the institution's capability at managing it successfully. At one extreme, commercial loans and residential mortgage loans clearly lie within the ambit of regulatory capabilities and hence of activities that should be permitted for depositories. At the other extreme, the operation of a delicatessen is surely an activity about which bank and thrift regulators are unlikely ever to have adequate expertise; it is an example of the type of activity that should not be allowed to be part of the insured depository, except perhaps, for a de minimus amount of unorthodox investments or activities that would be permitted to occur within the insured entity, provided that their aggregate potential for loss comes to only a small percentage—say 10 percent—of the depository's net worth).[40]

Under this criterion, the inherent riskiness of an investment or activity would not automatically disqualify it—so long as regulators could assess that risk, determine the appropriate net worth and premiums, and generally monitor it. Also, the argument, "insured deposits were not meant to be used to finance activity X....," should have little or no validity, once activity X has met the monitoring criterion. It is important to remember that commercial loans offered by banks, funded by insured deposits, are used to finance virtually every type of economic activity that occurs in the U.S. economy.

This monitoring criterion would mean a fundamental merging and uniformity of the powers of banks and thrifts. Depositories that choose to remain (or become) relatively specialized as residential mortgage lenders should probably remain with a separate regulator, however, because the problems of mortgage finance do require special regulatory expertise.[41] There could still be room for distinctions, though more limited than the past, between state-chartered and federally chartered institutions; federal deposit insurance and regulation could continue to be an overlay on state-regulated institutions. The states could continue to play a role as regulators and innovators; it is worth recalling that adjustable rate mortgages and NOW accounts were originally the innovations of state-chartered institutions. Where safety-and-soundness concerns would mean serious risks to the federal deposit insurer, however, the federal regime must have the ultimate overriding powers.

Further, banks and thrifts should have the ability to branch and conduct business throughout the United States, provided that their activities and expansion are otherwise consistent with safe-and-sound behavior and adequate net worth. The tradition of localism in U.S. banking, enshrined in the McFadden Act of 1927 and the Douglas Amendment to the Bank Holding Company Act of 1956, has created a fragmented industry, with far too many small and inefficient institutions. Banks and thrifts that could efficiently expand or merge into new markets have not been able to do so. Financial services customers have been deprived of the improved services and increased competition that this entry could bring. In addition, geographic diversification, if done prudently, could increase the safety and soundness of these institutions, since they would be less tied to the economic fortunes of a single locality or region and could offset losses in one area with the profits from others.[42]

The holding companies of depository institutions should be allowed to engage in any economic activities that are otherwise permissible to other companies in the U.S. economy. The approach of the Bank Holding Company Acts and the Glass-Steagall Act—to limit bank holding companies to financial activities that are closely related to banking—makes little economic sense.[43] The fears of abuses that might arise from wider ownership of banks—possible abuses of market power or of customer deception—are largely unfounded. If there are genuine efficiencies and synergies (economies of scope) from a nonfinancial company's owning a bank, that ownership should be welcomed, not avoided.

The experience of thrift holding companies is instructive. The presence of companies involved in markets as diverse as autos, steel, wood products, retailing, public utilities, insurance, and securities as holding company owners of thrifts has not created problems; the same would surely be true if these, or similar, companies had owned banks. State and federal antitrust laws and consumer protection laws should be adequate to deal with the occasional problem of market power or consumer deception abuse that might otherwise arise from wider bank holding company ownership.

There is, however, one set of issues related to owners, including holding companies (or their affiliates), that warrants special regulatory attention: the financial flows and transfers between an insured depository and its owners.[44] As a review of Table 11-1 indicates, the transfer or draining of even modest amounts of assets from the depository to its owner (or an affiliate of the owner) can place the deposit insurer at serious risk. Such transfers of assets can happen in a number of ways: (1) the depository can declare a large dividend or repurchase a large fraction of its stock, and thereby transfer cash or other assets to the owners; (2) the depository can lend to or invest in projects of the owner at preferential rates or without adequate security, or the depository could make the loans to the owner's friends, associates, customers, or suppliers; (3) the depository can overpay for items that it purchases from the owner (or the owner's customers, etc.); (4) the depository can undercharge for items that it sells to the owner.

If the depository's (market value) net worth levels are adequate and the dividends and other transactions (in aggregate) are small (relative to net worth), then these actions can cause little harm. But, as the depository's net worth grows thinner (or is not based on market value accounting) and the transactions grow larger, the possibilities for abuse that put the deposit insurer at risk also increase.

As a consequence, the dividends (or stock repurchases) of banks and thrifts should be limited to amounts that do not place the institution in jeopardy of falling below the minimum net worth requirements[45]; banks and thrifts that are not meeting their minimum net worth requirements should not be allowed to declare any dividends at all. Large transactions with owners (or their affiliates, customers, or suppliers) should be strongly discouraged, and *any* transactions with owners (or their affiliates, etc.) should be at prices and terms that reflect realistic market alternatives and can be documented to be so. Close regulatory scrutiny and vigilance is vital in this area.

Using Evidence to Support Regulations

The bank and thrift regulators, in their development of safety-and-soundness regulations, must make better use of the financial data that are regularly collected from their banks and thrifts. If necessary, they should supplement those data with special surveys or studies.

In virtually all instances, safety-and-soundness regulations in the bank and thrift area have been developed on the basis of the judgments of the personnel in the agencies, with little or no use of any data to support these judgments. This style of rulemaking would not be accepted in most other regulatory areas of the federal government. It should not be accepted here.

All of the bank and thrift regulatory agencies have trained economists, econometricians, and statisticians who can provide the appropriate statistical studies. More could readily be hired.

The bank and thrift trade associations and the larger institutions in the two industries should do their part, as well. They should be conducting their own statistical studies and submitting the results of these studies as part of their comments on proposed regulations. Such submissions would put more pressure on agencies to conduct their own studies. Better-grounded regulations would surely emerge from this process.

Education

The FIRREA instructs the regulatory agencies to conduct "training seminars in risk management" for their own employees and for the employees of insured institutions and to consider the possibility of a more formal program. The agencies should grasp this opportunity and expand upon it.

Insurance companies frequently have educational programs, conduct advertising, or have other means by which they inform their insureds about risks and encourage and work with the latter to reduce those risks. The bank and thrift regulators currently do a little of this, by speaking at trade association conventions and other meetings, sponsoring seminars, and conducting and disseminating research; much more, however, could and should be done.

All of the regulatory agencies currently have formal training programs for their examiners and supervisors, which cover financial theory and practice, accounting, appraising, and other topics that are relevant to understanding the operation of a modern depository institution. These programs should be opened to the senior managers of banks and thrifts, and the agencies should encourage those managers' attendance and participation. Certification programs could be developed. After the training programs have been in place for a while, the agencies should seriously consider a licensing program for senior managers. A reasonable parallel could be drawn with airplane pilot certification and licensing. If the regulatory agencies have the responsibility for maintaining the safety and soundness of depository institutions and of protecting the insurance funds, they should properly be concerned about the education and skills of the "pilots" and "copilots" of those institutions, in order to reduce the number and severity of the "accidents" that the latter may cause.

Are More Radical Reforms Necessary?

Though the reforms proposed in this chapter are quite extensive, there have been proposals for more radical reforms of the deposit insurance system. One group of proposals calls explicitly for the replacement of government deposit insurance with private deposit insurance. The rollbacks or limitations proposals mentioned previously might achieve this outcome. Other proposals in this vein include placing a cap on the aggregate amount of government-provided insurance at current levels,[46] thereby forcing any expansion of deposit insurance to occur through private insurers, or simply phasing out government insurance completely over a specified time period, such as ten years.[47]

Private insurance does not appear to be a viable solution for the special problems of deposits and depositories. The presence of private insurers would not solve the problem of the safety and liquidity of deposits; the crucial question would only be moved one step back: How can the safety and soundness of *private* deposit *insurers* be assured, in order to provide these features for the ultimate depositors?[48]

Further, the history of deposit insurance is instructive. During the 100 years prior to 1933 and the establishment of the FDIC and the FSLIC, there were no private insurers of deposits,[49] despite—or perhaps because of?—numerous bank failures and losses by depositors.[50] Even in the years after 1933, there have been no private insurers of the deposit amounts above the government-insured levels.[51]

Another set of proposals would narrow considerably the types of assets in which government-insured depositories could invest, in order to confine them to virtually risk-free assets, such as short-term Treasury bills or other short-term high quality debt obligations.[52] In essence, these "narrow banks" would look very much like today's money market mutual funds, except that the deposits would be formally insured and the narrow banks would have more depositor conveniences, such as more bricks-and-mortar locations, automated teller machines, and greater check-writing flexibility.

These narrow banks would be a radical means of achieving many of the reforms advocated in this chapter: market value accounting (since short-term Treasury bills would always trade very close to par); risk-related net worth standards and insurance premiums (since short-term Treasury bills carry little interest rate risk or credit risk); complete coverage of deposits; and a drastic change in the regulatory boundary of what type of assets can be included within the insured entity. Under this type of proposal, the more traditional types of lending (e.g., commercial loans, mortgage loans, consumer loans) that are currently performed by depositories would occur through non-insured entities: either free-standing companies, as occurs today through industrial credit companies, factoring companies, mortgage bankers, and consumer credit companies; or holding companies (or their separate subsidiaries) of the narrow banks. The funding for this lending would have to come from the issuance of liabilities (e.g., commercial paper, notes, bonds) by these free standing companies or holding companies (but not their narrow bank subsidiary) that were explicitly not government insured.

A modification of this narrow bank proposal would allow insured depositories to invest in any asset that could readily be marked to market on a *daily* basis.[53] In practice, this approach would allow depositories to invest in government securities, corporate stocks and bonds, commercial paper, and mortgage-backed securities. One should not rule out the possibility, however, that the implementation of a radical approach of this kind could lead to a greater standardization and securitization of other types of loans, thereby widening the range of allowable investments for these insured depositories.

These narrow bank proposals have considerable appeal because they would achieve many of the necessary reforms, and refinements would surely be possible. They would, however, cause the loss of the synergies and efficiencies that currently flow from combining the lending and deposit-taking functions within a single entity.[54] And they would involve a massive upheaval and reorganization of the U.S. system of depository institutions, on a scale unprecedented in U.S. history.[55]

The package of extensive but more gradualist reforms that were presented earlier in this chapter is a superior route. They would best achieve the vital goals of a stable and efficient financial services sector and the avoidance of future catastrophic losses to the deposit insurance funds. Their costs would be relatively modest, and the accompanying disruption to the current institutional framework for depositories would also be relatively modest. But the daily-market-price-of-assets version of the narrow bank proposal, because it does embody many of the important reforms, would be an acceptable alternative.

Notes

1. See Spellman (1982), Marcus (1984), and Keeley (1989).
2. See, for example, Brumbaugh and Litan (1990). Also, in 1989 the mutual savings banks (MSBs) in aggregate ran a modest loss, $97

million, after being profitable in 1988; a quarter of this group ran losses of $1.3 billion, which exceeded the $1.2 billion in profits by the remaining three-quarters.

3. Though the discussion in this chapter will focus primarily on banks and thrifts, the lessons for regulation and deposit insurance are equally applicable to credit unions.

4. For other writers' recommendations for reforming depository regulation and insurance, see Kane (1985; 1986; 1989), Benston et al. (1986), Litan (1987), Brumbaugh and Carron (1987), Boyd and Rolnick (1988), Brumbaugh (1988), Bernheim (1988), Benston and Kaufman (1988a; 1988b), Scott (1989), and Benston et al. (1989).

5. This section draws heavily on White (1988a; 1988c; 1990a; 1990c; 1991); see also Benston (1989), Berger, Kuester, and O'Brien (1989), Brewer (1989), and Simonson and Hempel (1989).

6. Less any amortization.

7. This type of systematic behavior is usually described as *gains trading*.

8. Even if the markets are thin as compared with the amount of assets to be priced, those price quotations would almost always provide a better estimate of current value than would the historical cost of the assets.

9. That this is feasible is indicated by the Department of Labor's requirement (under the Employee Retirement Income Security Act of 1974) that pension fund managers must periodically mark their real estate holdings to market through appraisals.

10. See, for example, Berger, Kuester, and O'Brien (1989) and Engelke (1990).

11. Increasingly, GAAP itself is relying on *estimates* of the values of some assets and liabilities.

12. There is adequate precedent for this confidential reporting to regulators. From 1984 through 1989, thrifts reported interest rate risk information only to the Federal Home Loan Bank Board but not to the public.

13. See Kane (1985; 1988; 1989).

14. This has been true for state-sponsored insurance funds as well as the federal insurance funds; see Kane (1988; 1989).

15. In early 1990, pleas for regulatory forbearance for commercial banks began to mount.

16. One of the FIRREA's unfortunate side effects was the short-term *elimination* of an interest rate risk component to the net worth standards that apply to thrifts. Between 1987 and 1989 the thrifts' net worth standards included an interest rate risk component. By forcing the OTS to issue new risk-based net worth standards in November 1989, before the agency had completed its preparations of a new interest rate risk approach (the former approach could not be included in the new

standards), the FIRREA created a serious short-term hole in those standards. As of the summer of 1990, it is unclear how soon the OTS will be able to develop and promulgate its interest rate risk component so as to integrate that component into the overall risk-based net worth requirements.

17. In the bank regulators' risk-based net worth standards, commercial loans require 8 percent net worth (as a percentage of asset value), whereas residential mortgage loans require only 4 percent net worth, Fannie Mae and Freddie Mac mortgage-backed securities require only 1.6 percent net worth, and Ginnie Mae mortgage-backed securities require 0 percent net worth.

18. See, for example, Benston and Kaufman (1988a; 1988b) and Eisenbeis (1990).

19. This argument is made in different ways by Chan, Greenbaum, and Thakor (1988) and Flannery (1989).

20. See White and Golding (1989).

21. Similarly, if the holders of banks' overseas deposits receive protection from the deposit insurance fund, premiums should be levied on them as well.

22. See Horvitz (1983a), Goodman and Shaffer (1984), and Benston and Kaufman (1988a; 1988b).

23. It is sometimes argued that the riskiness of some activities, such as loans to Third World countries or loans on commercial real estate in the Southwest, can only be discovered after the fact, when higher premiums (or higher net worth standards) for those activities would have little effect. This argument is correct with respect to the loans that have already been made, but the incentives are important for any new loans that might be made.

24. In the terminology of Chapter 3, it is the exercise of the regulators' call options.

25. If a single, identifiable acquirer cannot be found, the regulator could declare a receivership, eliminate the former owners and become the de facto new owner, inject the necessary funds to bring the institution to an adequate level of net worth (on a market value basis), install new management, and then sell its ownership shares to the public. This is a stylized version of the FDIC's actions when it had to deal with Continental Illinois in 1984; see Sprague (1986).

26. This would include the interests of subordinated debt holders.

27. Friday afternoon closures, followed by Monday openings under new auspices, however, would seem to provide the necessary continuity.

28. In late 1988 the Bank Board issued an advance notice of proposed rulemaking along these lines. Also, in a similar spirit, in 1989 the OCC indicated that in determinations of solvency it would no longer allow loan loss reserves to be added back to a bank's assets.

29. In the language of Chapter 3, they have only the value of the unexercised put option.

30. One cannot rule out the possibility that a bank or thrift, prior to the appointment of a receiver, has a small but positive market value net worth, which is then dissipated by the insurance fund's handling of the disposal of the institution. Such instances, however, are likely to be exceedingly rare. All of the available evidence indicates that market values are substantially below the reported book values of banks and thrifts that are approaching insolvency; see, for example, Barth, Brumbaugh, Sauerhaft, and Wang (1989).

31. Whether it should be extended to foreign deposits would depend on whether those deposits are the obligation of the domestic U.S. depository or whether they are the obligation of the holding company. If the latter is the case, they should not be covered. If the former is the case, then they should be covered, since they could generate the runs problem that is a major justification for deposit insurance. If they are covered, then they should be part of the premium base.

32. Also, the time span that would apply to these limitations per individual is usually not specified. Should the limitation apply over that person's entire lifetime? Or would the limitation be renewable at some point in time after any exhausting of the allowable insured amount? How soon?

33. The FIRREA does require the public announcement of final enforcement orders. With 100 percent coverage, however, much more could be made public.

34. See Horvitz (1982; 1983b; 1984), Baer (1985), Benston, et al. (1986, Ch. 7), and Bernheim (1988).

35. The interests of the subordinated debt holders are not identical to those of the deposit insurer because the former's losses are limited to their investments, whereas the latter bears all losses up to the value of the insured depositors' claims.

36. They might require some modifications for a market value accounting framework.

37. An exception might be made for small banks and thrifts that might not be able to issue enough subordinated debt in aggregate to warrant the attention of institutional holders. Stringent disclosure requirements should accompany any small denomination issues.

38. The Bank Board's net worth requirements for thrifts allowed subordinated debt to count fully as net worth until seven years before its maturity, at which time its value for net worth calculations was reduced by one-seventh per year.

39. With little owners' equity at stake, the risk-taking incentives of the owners might overwhelm the protections created by the subordinated debt holders.

40. The placement of these unorthodox activities in a subsidiary of the

insured institution does not protect the insurer since the loss of value in the subsidiary would reduce the value of the institution's ownership interest in that subsidiary.

41. Separate regulatory agencies would give more room for creativity and independent development of new approaches to regulatory problems; see Benston (1984) and Fischel, Rosenfeld, and Stillman (1987). The same argument could be made for maintaining a separate insurance fund for thrifts; alas, that ship has sailed and is unlikely to return.

42. See McCall and Lane (1980), U.S. Department of the Treasury (1981), E. White (1983), and Ogden, Rangan, and Stanley (1989).

43. See Walter (1985), Litan (1987), E. White (1989), and Benston (1989).

44. This is currently treated under a set of regulations that are usually referred to as *transactions with affiliates.*

45. A premium that is geared to net worth should also provide a disincentive to such dividends.

46. See Smith and Tammen (1989), and Tammen (1990).

47. See England (1989).

48. See Baer (1985).

49. There were, however, bank clearinghouses in the nineteenth century in some cities; see Gorton (1989). But the clearinghouses did not provide complete insurance and did not stop all bank runs.

50. See Rolnick and Weber (1983), E. White (1983), and Calomiris (1989a; 1989b).

51. There have been, though, state-sponsored insurance plans for state-chartered thrifts; see Kane (1988; 1989).

52. See Litan (1987).

53. See Guttentag and Herring (1988) and Benston et al. (1989).

54. See Benston and Kaufman (1988a; 1988b).

55. Also, the uninsured liabilities of the insured depositories' holding companies and affiliates might attain implicit insured status.

CHAPTER TWELVE

Conclusion

It is difficult to write a conclusion to a book about the S&L debacle.

At one level, an important aspect of the S&L debacle has indeed concluded. As Chapter 7 indicated, the abusive actions by hundreds of thrifts in the mid-1980s have come to an end, although much of the costs of those actions are still being freshly recognized in the 1990s. Though the specific confluence of events that produced the S&L debacle of the 1980s are unlikely to recur, fundamental problems with thrift *and bank* regulation and deposit insurance remain and have not been repaired by the cleanup legislation of 1989. Substantial reforms of bank and thrift regulation and deposit insurance, along the lines advocated in Chapter 11, are vital for encouraging an efficient and flexible system of depository institutions, while also preventing new abuses and losses to the deposit insurance funds.

At another level, the complete saga of the S&L debacle, with its full ramifications and consequences, is not yet over. Updated estimates of the costs and numbers of failed thrifts will be headline news for many years to come. The Congress will likely have to re-address the costs of the cleanup in 1991 or 1992, and the general reform of depository regulation and insurance will be on the legislative agenda at the same time.

In addition, the technologies that underlie financial services—telecommunications and data processing—will continue to change and improve, with significant consequences for banks and thrifts. These improving technologies will mean a continued erosion of geographic and product-line barriers to competition; the securitization of assets, discussed later in the chapter, is one major facet of this change. The environment for banks and thrifts is sure to become more competitive, bringing further changes and adaptations. These changes will pose new challenges for depository regulation and insurance. The 1990s will be a decade of flux in financial services.

Nevertheless, a large number of important lessons can be learned from the 1980s experience of the thrift debacle. The extraction of these lessons from the debris of the debacle has been one of the major goals of this book. These lessons have broad relevance for the future of bank and thrift regulation and deposit insurance in the United States—and in other countries that wish to avoid many of the costly mistakes that the U.S. Government committed during the 1980s.

A review of these lessons will constitute the final section of this chapter. First, though, two interrelated themes that highlight the relevance and importance of those lessons will be explored: the effects of technological change in increasing competition in the bank and thrift sectors, and the future of the thrift industry in the financial services markets.

Technological Change and Increasing Competition

Financial services are an information-intensive business. Financial services firms have to gather information on actual and potential customers (e.g., borrowers), process it, verify it, evaluate it, and store it for possible future retrieval. After loans are made, interest and principal payments must be received and processed, and the general condition of the borrower and the loan periodically monitored. At the heart of these processes are the technologies of telecommunications and computerized data processing.

As these technologies improve and the costs of these information-intensive transactions decrease, existing financial services firms can expand their geographic coverage since lower telecommunications costs mean that information can be consistently and effectively gathered from farther afield, and they can extend the scope of their product and services offerings; new firms can take advantage of advances in information processing and offer new services in new market niches.

The implications for banks and thrifts are direct. As the technologies continue to improve, banks and thrifts will find it easier to offer their services over broader geographical areas. As states reduce their barriers to intra- and interstate branching, some banks and thrifts will surely find expansion worthwhile and will enter each others' "home" territories. Entry by overseas banks into American markets is also likely to increase. The result can only be greater competition in these markets.

Further, nondepository firms will continue to encroach on the markets of banks and thrifts. On the liabilities (deposits) side of banks' and thrifts' balance sheets, money market mutual funds will continue to provide a close substitute that is attractive to some depositors. Other types of mutual funds, pension funds, and insurance companies, as well as direct investments in securities, are more distant substitutes for deposits; but they do provide some competition with banks and thrifts for depositors, and this competition could well become sharper.

On the assets side, all of the financial intermediaries listed at the beginning of Chapter 10 are currently competitors for some types of bank and thrift lending, and their competition is likely to become more intense. Changes in regulations (e.g., registration of securities) and in tax laws (e.g., the tax deductibility of interest payments) could lead to more direct issuance of securities that are competitive with borrowings from banks and thrifts. Expanded corporate use of short-term borrowings through the commercial paper market (to replace short-term borrowings from banks) and of longer term borrowings through the "junk bond" market (to replace longer term borrowings from banks) are examples from the past.

Securitization—the combining of loans into packages and the issuance of securities representing claims on the interest and repayment of principal on those loans—are another means whereby competition will intensify for banks and thrifts. This process has proceeded furthest in the area of home mortgages, and the securitization in that area can serve as a good example of how the process intensifies competition.

Table 12-1 illustrates the difference between the traditional portfolio lender approach to home mortgages and the securitization approach. The traditional portfolio approach involves vertical integration within a thrift (or a bank) of most of the essential steps of the financing of a home mortgage: A thrift (or bank) originates the mortgage (makes the actual loan); services it (makes sure the payments are made in a timely manner and attempts to correct delinquencies); collects the savings from the ultimate funders (the depositors); and issues the deposit liabilities to those funders. The FSLIC and the FDIC have been the government guarantors of those deposit liabilities.

In the securitized version, these individual processes are no longer as vertically integrated. A mortgage banker (as well as banks and thrifts, to the extent they participate in the securitized version— i.e., to the extent that they act as a mortgage banker) can originate and service the mortgage[1]; but the mortgage banker does not hold the loan in its portfolio. Instead, it sells the loan to Freddie Mac, Fannie Mae, Ginnie Mae, or an investment banker.[2] These "conduits" package the mortgage loans and sell mortgage-backed securities (liabilities), which represent claims on the interest and principal repayments

Table 12-1
The Stages in Originating and Funding Residential
Mortgages: The Traditional Portfolio Lender
Approach and the Securitization Approach

Stages	Traditional Portfolio Lender Approach	Securitization Approach
Originate mortgages	Thrifts Banks	Mortgage bankers Thrifts Banks
Service mortgages	Thrifts Banks	Mortgage bankers Thrifts Banks
Collect savings for funding mortgages	Thrifts Banks Deposit brokers	Insurance companies Pension funds Mutual funds Individual investors Thrifts Banks Deposit brokers
Issue deposits or liabilities to funders	Thrifts Banks	Freddie Mac Fannie Mae Ginnie Mae Investment bankers
Guarantee liabilities	FSLIC FDIC	Freddie Mac Fannie Mae Ginnie Mae

Mortgage-Backed Security

of these loans.[3] The buyers of these MBSs—insurance companies, pension funds, mutual funds, and other investors[4]—are the gatherers of the savings that are providing the funding for these mortgages; and Ginnie Mae (as a full-fledged government entity) and Freddie Mac and Fannie Mae (as Congressionally created, though privately owned, entities) guarantee the liabilities. As of year-end 1988, these three agencies had $789 billion in MBSs outstanding; this was 36.5 percent of the total of all home mortgage loans outstanding at this date.[5]

The securitization method has two effects on the mortgage loan market. First, it primarily provides new competition for the traditional portfolio lenders. The beneficiaries of insurance companies, pension funds, and the other buyers of the MBSs can participate in the earnings streams derived from mortgages without directly or indirectly depositing their funds in a thrift or bank, thanks to the mortgage bankers who originate those mortgages and the conduits who buy them and issue the MBSs. The entire process takes place outside of and in competition with the traditional portfolio lenders. Second, it does provide a liquid secondary market for mortgages and thus make it easier for a portfolio lender to convert mortgages into cash as desired. Unless a thrift or bank wants to behave mostly as a mortgage banker, however, the first effect dominates the second and causes competition in the mortgage market to be more intense than it otherwise would be.

Securitization is less well developed in other lending areas; it does exist, though, for auto loans, credit card loans, computer leases, and student loans.[6] Similar to home mortgages, securitization allows nonbank entities (e.g., automobile companies, finance companies, retail chains) to originate loans and sell them. It also allows funders to obtain claims on the earnings streams from these loans without placing their funds in a bank. Except for student loans (where a government-created but privately owned entity, the Student Loan Marketing Association, or Sallie Mae, is the conduit), government or government-sponsored agencies have not been present as a driving force. But the pattern of standardizing loans, packaging them, and issuing securities is now well established, and this securitization process will surely expand in future years.

In, sum, continuing improvements in technologies can only mean greater competition in the markets traditionally served by banks and thrifts. This prospect of increased competition makes the absorption of the lessons of the 1980s thrift experience and the adoption of the fundamental reforms of Chapter 11 yet more crucial.

Do Thrifts Have a Future?

No book can be written about the S&L debacle without addressing the future of the thrift industry. This question is frequently asked in terms of, "Do we still need a thrift industry?" This specific version of the question appears to be asking whether a separate group of depository institutions should be cordoned off and required to specialize in home mortgage finance, in return for special favors and coddling.

This is the wrong question to ask. "Need" is not the right way to approach the question of the future of the thrift industry. It is now clear that we do not "need" the thrift industry. There are other ways to deliver mortgage finance to the marketplace. The securitization process just described is one; commercial banks and credit unions are another; the investment banking industry could likely develop yet others. Indeed, the same answer should be given if the question were asked, "Do we still *need* commercial banks?" As Table 10-1 made clear, there are many other sources of lending in the U.S. economy.

At the same time, however, the recognition that borrowers do not "need" the thrift (or banking) industry is not an excuse for needlessly or thoughtlessly taxing and/or regulating them to extinction.

Instead, a different set of questions should be asked: "Is there a group of entrepreneurs and managers whose skills are best focused on originating and holding residential mortgages and gathering deposits, with modest diversification into related financial services? Further, can these specialist lenders survive in an environment of ever-stronger competition and an enhanced concern for safety-and-soundness regulation and the protection of the deposit insurance funds? Are consumers, as well as taxpayers, better off because of the presence of these lenders in the marketplace?"

The answers to these questions are "yes."[7] This group of efficient thrifts is definitely smaller than the current population of thrifts. Substantial shrinkage will surely occur in their numbers and, at least for a while, in their aggregate assets. But, even in the adverse economic operating environment (for thrifts) of 1989, more than 1,800 thrifts (with over $750 billion in assets) were profitable.

It is also worth recalling that the advent of the adjustable rate mortgage (ARM) and other hedging methods have allowed thrifts greatly to reduce their interest rate mismatch. ARMs now typically account for anywhere from 40 percent to 70 percent of thrifts' mortgage originations on a month-to-month basis; as of year-end 1988 ARMs accounted for 55 percent of the mortgages held by thrifts.

These figures do not imply, however, that interest rate risk nirvana has been reached. In aggregate, thrifts are still mismatched and harmed by increases in interest rates. Much more needs to be done, through appropriate risk-based net worth standards and deposit insurance premiums, to induce thrifts to continue to reduce their interest rate risk or to have enough net worth to bear that risk.[8] Still, much progress has been made since the late 1970s.

Intense competition in the mortgage markets will continue to be the environment that these efficient thrifts will face, and many of the marginally profitable thrifts may well falter and disappear. The punitive effects of the FIRREA (see Chapter 9) in requiring higher insurance premiums and levies on the Federal Home Loan Banks, in mandating reduced operating flexibility, and in creating greater uncertainty generally about the future of thrifts will not make their financial operations any easier. On the other hand, as the insolvents are cleared from the scene, the operating environment of the remaining thrifts should improve, and their profit margins should widen. Further, there does seem to be room for cost reductions as a strategy for profit improvements. As Table 12-2 indicates, operating

Table 12-2
Operating Expenses as a Percentage of Assets,
FSLIC-Insured Thrifts, 1960–1989

	Operating Expenses, as a Percentage of Assets[a]
1960	1.14%
1965	1.06
1970	1.11
1975	1.20
1980	1.28
1985	1.83
1986	1.97
1987	1.93
1988	1.77
1989[b]	2.05

[a] Assets as of end of year.

[b] SAIF-insured thrifts, not in RTC conservatorship.

Sources: FHLBB and OTS data.

expenses as a percentage of assets widened considerably in the 1980s, as compared with previous years. At least part of this widening was due to the new activities permitted for thrifts in the 1980s, but there may well have been a bloating of costs that accompanied the general expansion and enthusiasm of the mid-1980s. Careful attention to cost containment and cost reduction could well yield high substantial benefits for thrifts.

Ought these efficient thrifts to be specially subsidized, protected, or otherwise favored by government policies? The answer must be in the negative. But should they be needlessly eliminated from the marketplace by excessive taxation or unduly restrictive regulation? The answer here must also be negative. To eliminate them in this way would be a serious waste of the human, physical, and reputational resources that are currently invested in these efficient firms. They should neither be forced to specialize (if their skills are better devoted to prudent diversification) nor to diversify (if their skills are better devoted to specialization). Instead, they should be subject to the same regulatory and financial conditions as other depositories and thereby allowed the flexibility to find the mix of activities that best suits their talents, while avoiding the placement of undue risk on the deposit insurer.

It appears that the financial services markets are much like retailing in general. In retailing, the market appears to support a variety of store types, corresponding to the skills of their proprietors: single store specialty boutiques, chains of specialty stores, full-line department stores of various sizes, broad-based supermarkets of various sizes, and so on. The same appears to be true for banks and thrifts. Some will be good at being specialists; others will be better at being "department stores." Some can remain at modest size and pay close attention to costs and to the demands of their customers; others can take advantage of the technologies discussed earlier in this chapter and expand in size, product scope, and geographic domain. Public policy, while ensuring safety and soundness, should give banks *and thrifts* the flexibility to pursue the strategies that they find most efficient. The package of reforms presented in Chapter 11 would constitute this policy.

In sum, there do seem to be a group of efficient depositories who are currently specialists in residential mortgage finance and who can survive in a competitive marketplace, but who also need the flexibility to adapt to changing circumstances. Public policy should cease trying to force them to specialize—either accompanied by coddling and special favors, as was true before 1980, or by punishment and taxes, as was true after 1989—and should instead treat them on an equal footing with other depositories. Many will choose to remain as mortgage finance specialists, while attaching a few complementary services; others will choose to diversify. The executives of the former group will probably still think of themselves as operators of "thrifts," will have trade groups that are sensitive to their special problems, and (as Chapter 11 argued) are probably best regulated by an agency that has the expertise to deal with their problems. *The basic choices, however, should be determined by efficiency in the marketplace, not by legislative fiat.*

Reviewing the Lessons of the Debacle

The basic lessons to be learned from the S&L debacle of the 1980s are important and warrant a concluding review.

1. Deposit insurance can and should be treated like other forms of insurance. From the perspective of the deposit insurer, the net worth of a bank or thrift acts like a deductible in protecting the insurer against accidents, carelessness, or deliberate risk-taking. But that net worth must be measured on a current market value accounting basis, not on a backward-looking cost-based accounting framework. Risk-based premiums are a natural complement to net worth standards in protecting the insurer, as is the expanded use of long-term subordinated debt. (Chapters 3, 10, 11)

2. The policy efforts of the 1960s and 1970s to preserve and protect the thrift industry as a narrowly focused lender of long-term fixed-rate mortgages were serious errors. The application of the Regulation Q deposit interest ceilings to the thrifts in 1966 and the Congress's hostility to adjustable rate mortgages in the 1970s prevented thrifts from adapting and becoming more flexible and diversified lenders and exacerbated the effects on them of the sharply higher interest rates of the late 1970s and early 1980s. (Chapter 4)

3. The economic deregulation of the early 1980s was and remains a fundamentally sensible approach; but it happened fifteen years too late. The thrift industry could not then and cannot now remain an anachronistic part of the financial services industry, frozen in the 1950s and 1960s, while the increasingly competitive financial markets of the 1980s and 1990s pass it by. But that deregulation approach, set in the economic environment of the early 1980s, created an over-

whelming set of opportunities, capabilities, and incentives for risk-taking. This economic deregulation needed desperately to be accompanied by stepped-up safety-and-soundness regulation and increased attention to creating economic incentives to reduce risk-taking. Instead, perverse federal actions actually weakened the safety-and-soundness regulatory system; and the 1980s changes in the nation's tax laws and a substantial drop in the price of oil then compounded what was already destined to be a difficult and costly situation. (Chapters 5 and 6)

4. Hundreds of savings and loans took advantage of these opportunities, capabilities, and incentives and expanded rapidly between year-end 1982 and year-end 1985; much of their expansion was in new types of lending and investments. In too many instances these thrifts' executives were overly optimistic, excessively aggressive, careless, ignorant, and/or outright criminal or fraudulent. Their gambles did not succeed, and these thrifts fell into insolvency, generating eventual huge costs for the government deposit insurer, the FSLIC. (Chapters 6 and 7)

5. The majority of the thrift industry did not indulge in the abuses of their eventually insolvent namesakes and have remained largely healthy. They have, however, been tarnished politically by their industry's lobbying practices and harmed financially by the uneconomic pricing and practices of their insolvent competitors, by the public's perception of all thrifts as financially suspect, and by some of the punishing aspects of the Financial Institutions Reform, Recovery, and Enforcement Act of 1989. (Chapters 6 and 9)

6. The Federal Home Loan Bank Board's tightening of its regulations during 1985-1987 brought the growth binge and the abuses largely to a halt, but three years too late. Unfortunately, the bad loans and investments that had already become part of these thrifts' portfolios could not be cured or reversed by these tightened regulations. Partly because of a poor and backward-looking information (accounting) system and partly because the real estate markets of the Southwest and elsewhere began to crumble only after 1985, the revelation of the losses embedded in these assets began only in 1986 and accelerated in the following years. The losses were largely the writedowns of the values of these assets, but the losses also included operating losses that were the product of the inadequate (or zero) income being earned on these poorly conceived loans and investments and the higher deposit interest costs arising from depositors' ner-

vousness about the safety of their deposits in these insolvent or soon-to-be insolvent thrifts. (Chapters 7 and 8)

7. The insolvencies of these hundreds of thrifts triggered the obligations of the deposit insurer, the FSLIC, to keep the insured depositors whole. The obligations of the FSLIC were approximately equal to the difference between the inadequate market values of these thrifts' assets and the larger values of their deposit liabilities. Once these thrifts had made these shoddy loans and investments, the FSLIC's obligations were largely unavoidable. But delay in disposing of the insolvents, as occurred from 1985 on, could only make these costs larger. Though supervisory controls were usually in place to prevent further abuses, the occasional thrift might slip past those controls and engage in more risk-taking; and even with effective controls, the "caretaker" managers of insolvent thrifts had few incentives to manage assets and liabilities carefully in order to avoid further deterioration and erosion of values. More important, the flawed and weakened accounting system allowed many thrifts that were insolvent on a market value basis to report solvency or even a healthy net worth and thereby to continue to expand and take risks. (Chapter 7)

8. The delay in the enactment of the Competitive Equality Banking Act of 1987 was moderately costly, but earlier enactment with larger sums would not have reduced the costs of the debacle by much. The vast majority of the bad loans had already been made by mid-1986, and much of the remainder would not have been stopped by the CEBA; the huge losses were already embedded but not yet recognized. (Chapter 7)

9. The FSLIC's disposals of 205 insolvent thrifts in 1988 were an important and necessary start on the huge backlog of insolvent thrifts. These disposals—mostly through placements with acquirers— (1) were less costly than the alternative of eventual liquidation, (2) avoided "leaving too much money on the table" for the acquirers, and (3) made efficient use of the tax incentives that the Congress had provided. These transactions in aggregate were cost-effective means of stopping the hemorrhaging of these insolvents and reducing the government's eventual costs. (Chapter 8)

10. The expenditures of the large sums that have been necessary to dispose of the insolvents—either by liquidation and direct payouts to depositors, or by placements with acquirers and payments to them to take on the obligations to the depositors—were not and are not

"bailouts" in any meaningful sense of the word. The owners and managers of the insolvent thrifts have not received any of these moneys, and uninsured depositors have not been a significant presence in these insolvent thrifts. These payments are the direct (liquidation and payout) or indirect (placement with acquirers) honoring of the insurer's obligations—the U.S. government's obligations—to depositors. No one is being "bailed out." (Chapter 8 and 9)

11. The moneys that were lost on the insolvent thrifts' ill-conceived loans and investments are not recoverable to any significant extent. These losses are equal to the declines in values on these assets. Though suits and indictments should be brought against culpable executives and owners, these suits will yield little in recoverable moneys. (Chapter 6)

12. The Financial Institutions Reform, Recovery, and Enforcement Act of 1989 provided vitally needed financing for the continuation of the cleanup, and the Act's endorsement of higher net worth standards was a welcome recognition of the importance of net worth (and a reversal of the legislative attitude of a decade earlier). But the FIRREA did not provide enough funding, nor did it establish a permanent funding mechanism for dealing with deposit insurance fund problems; it missed the opportunity to embrace other fundamental reforms; it eliminated one major regulatory agency and replaced it with new agencies and relationships, thereby slowing the cleanup and raising its costs; and, as an expression of Congressional and Bush administration anger at the thrift industry, it taxed the remaining healthy thrifts, reduced their operating flexibility, and raised the distinct possibility that future legislation would tax and restrict them yet further—causing marginal thrifts to falter more quickly into insolvency and discouraging potential acquirers of insolvent (or on-the-edge) thrifts, and thereby raising the costs of the cleanup even higher. (Chapter 9)

13. The Resolution Trust Corporation must move quickly to clean up the remaining backlog of insolvent thrifts. The RTC must also begin to sell its real estate and other assets with all deliberate speed. Delay in the disposal of insolvent thrifts and delay of individual assets is costly. (Chapter 9)

14. The Congress must appropriate the necessary funds to allow the RTC to clean up the remaining insolvent thrifts. This spending is not discretionary in any meaningful sense. The funds are necessary to honor the U.S. government's insurance promises to depositors. A

refusal to spend the required moneys would mean a reneging on the government's obligations. (Chapters 9 and 11)

15. New legislation will likely be required to allocate more funds to the cleanup. Sooner is better than later. The Bush administration and the Congress must acknowledge this reality quickly; delay is costly. Further, the expenditures should not be subject to the discretionary appropriations process; the Treasury must be given an open-ended authorization to spend the necessary moneys to honor the federal government's insurance obligations. Along the way, the administration and the Congress should avoid increasing the taxation of (and the regulatory constrictions on) the healthy and efficient remainder of the thrift industry; and, if possible, they should give a clear signal as soon as possible that they intend to eschew such taxation, so that potential acquirers of insolvents will be attracted to the RTC's bargaining table and the costs of the cleanup can be reduced. (Chapters 9 and 11)

16. In order to prevent future debacles anywhere in the depository sector and to promote an efficient and flexible system of banks and thrifts for the 1990s and beyond, the Congress and the regulatory agencies must embrace a package of fundamental reforms: market value accounting; improved net worth (capital) standards based on market value accounting; risk-based premiums; prompt disposal of insolvents and early intervention; 100 percent insurance coverage of deposits; expanded use of long-term subordinated debt; automatic designation of the Treasury as the backup for the insurance funds; a reassessment of what activities can occur within an insured institution, a dropping of the restrictions on the activities that can occur in the holding companies of depositories, but continued tight scrutiny of the financial transactions between depositories and their owners (holding companies or otherwise); better use of the available data in the development of regulations; and improved financial education of the senior managers of banks and thrifts. (Chapters 10 and 11)

These lessons have been painful and costly. The S&L debacle has been a searing experience for all concerned. But the lessons must be learned; a failure to absorb them could be even more costly. They provide the basis for the improved public policy that is vital for a safe and sound system of efficient banks and thrifts for the 1990s and beyond.

Notes

1. The mortgage banker that originates the mortgages need not be the party that services it. The rights to service mortgage portfolios are actively bought and sold among mortgage bankers, thrifts, and commercial banks.
2. There are limits on the size of mortgage loan that Freddie Mac or Fannie Mae can buy, linked to average home prices nationwide; as of late 1989, this "conforming loan" limit was $187,450. The limits on the loans guaranteed by the FHA or the VA, which are the loans bought and packaged by Ginnie Mae, are substantially lower. The investment bankers tend to focus their securitization activities on the loans that are larger than these limits.
3. During the 1980s the types of MBSs issued became increasingly differentiated and sophisticated, so as to appeal to different types of investors; for an overview, see Campbell (1988, Ch. 19).
4. Banks and thrifts are also significant buyers of MBSs; as of the end of 1988, banks held $97 billion, or 15 percent of all MBSs; thrifts held $190 billion, or 29 percent; see Barth and Freund (1989).
5. See Barth and Freund (1989).
6. See Campbell (1988, Ch. 16).
7. See also Brumbaugh and Carron (1989).
8. Also, many ARMs are not sufficiently sensitive to interest rate increases and may not provide as much protection to thrifts as might appear at first glance.

Bibliography

Adams, Dirk S., and Rodney R. Peck, "The Federal Home Loan Banks and the Home Finance System," *Business Lawyer*, 43 (May 1988), pp. 833–864.

Adams, James R., *The Big Fix: Inside the S&L Scandal.* New York: John Wiley & Sons, 1990.

Baer, Herbert, "Private Prices, Public Insurance: The Price of Federal Deposit Insurance," *Economic Perspectives*, Federal Reserve Bank of Chicago, 9 (September/October 1985), pp. 45–57.

Bailey, Elizabeth E., Daniel R. Graham, and Daniel P. Kaplan, *Deregulating the Airlines.* Cambridge, Mass.: MIT Press, 1985.

Balbirer, Sheldon D., G. Donald Jud, and Frederick W. Lindall, "Regulation, Competition, and Abnormal Returns in the Market for Failed Thrifts," mimeo, University of North Carolina, February 1989.

Barth, James R., Philip F. Bartholomew, and Michael G. Bradley, "Reforming Federal Deposit Insurance: What Can Be Learned from Private Insurance Practices? " Reserve Paper No. 161, Office of Policy and Economic Research, Federal Home Loan Bank Board, June 1989.

Barth, James R., Philip F. Bartholomew, and Michael G. Bradley, "The Determinants of Thrift Resolution Costs," Research Paper no. 89–03, Office of the Chief Economist, Office of Thrift Supervision, November 1989.

Barth, James R., Philip F. Bartholomew, and Peter J. Elmer, "The Cost of Liquidating versus Selling Failed Thrift Institutions," Research Paper no. 89–02, Office of the Chief Economist, Office of Thrift Supervision, November 1989.

Barth, James R., Philip F. Bartholomew, and Carol J. Labich, "Moral Hazard and the Thrift Crisis: An Analysis of 1988 Resolutions," in Federal Reserve Bank of Chicago, *Banking System Risk: Charting a New Course.* Chicago: 1989, pp. 344–384.

Barth, James R., and Michael G. Bradley, "Thrift Deregulation and Federal Deposit Insurance, *Journal of Financial Services Research*, 2 (September 1989), pp. 231–259.

Barth, James R., Michael G. Bradley, and John J. Feid, "The Federal Deposit Insurance System: Origins and Omissions," Research Paper no. 153, Office of Policy and Economic Research, Federal Home Loan Bank Board, January 1989.

Barth, James R., Dan Brumbaugh, Daniel Sauerhaft, and George H. K. Wang, "Thrift-Institution Failures: Estimating the Regulator's Closure Rule," in George G. Kaufman, ed., *Research in Financial Services*, Volume 1. Greenwich, Conn.: JAI Press, 1989, pp 1–23.

Barth, James R., John J. Feid, Gabriel Riedel, and M. Hampton Tunis, "Alternative Federal Deposit Insurance Regimes," Research Paper no. 152, Office of Policy and Economic Research, Federal Home Loan Bank Board, January 1989.

Barth, James R., and James L. Freund, "The Evolving Financial Services Sector, 1970–1988," mimeo, Office of Thrift Supervision, September 1989.

Benston, George J., "Brokered Deposits and Deposit Insurance Reform," *Issues in Bank Regulation*, 8 (Spring 1984), pp. 17–24.

Benston, George J., *An Analysis of the Causes of Savings and Loan Association Failures*. Monograph Series in Finance and Economics no. 1985–4/5, Salomon Brothers Center for the Study of Financial Institutions, Graduate School of Business Administration, New York University, 1985.

Benston, George J., "Market Value Accounting: Benefits, Costs, and Incentives," in Federal Reserve Bank of Chicago, *Banking System Risk: Charting a New Course*. Chicago: 1989a, pp. 547–563.

Benston, George J., "Direct Investments and FSLIC Losses," in George G. Kaufman, ed., *Research in Financial Services*, Volume 1. Greenwich, Conn.: JAI Press, 1989b, pp. 25–77.

Benston, George J., "The Federal 'Safety Net' and the Repeal of the Glass-Steagall's Separation of Commercial and Investment Banking," *Journal of Financial Services Research*, 2 (September 1989c), pp. 287–305.

Benston, George J., R. Dan Brumbaugh, Jr., Jack M. Guttentag, Richard J. Herring, George G. Kaufman, Robert E. Litan, and Kenneth E. Scott, *Blueprint for Restructuring America's Financial Institutions: Report of a Task Force*. Washington, D.C.: Brookings Institution, 1989.

Benston, George J., Robert A. Eisenbeis, Paul M. Horvitz, Edward J. Kane, and George G. Kaufman, *Perspectives on Safe and Sound Banking: Past, Present and Future*. Cambridge, Mass.: MIT Press, 1986.

Benston, George J., and George G. Kaufman, *Risk and Solvency Regulation of Depository Institutions: Past Policies and Current Options*. Monograph Series in Finance and Economics no. 1988–1, Salomon Brothers Center for the Study of Financial Institutions, Graduate School of Business Administration, New York University, 1988a.

Benston, George J., and George G. Kaufman, "Regulating Bank Safety and Performance," in William S. Haraf and Rose Marie Kushmeider, eds., *Restructuring Banking and Financial Services in America*. Washington, D.C.: American Enterprise Institute, 1988b, pp. 63–99.

Benston, George J., and Michael Koehn, "Capital Dissipation, Deregulation, and the Insolvency of Thrifts," mimeo, Emory University, 1988.

Berger, Allen N., Kathleen A. Kuester, and James M. O'Brien, "Some Red Flags

Concerning Market Value Accounting," in Federal Reserve Bank of Chicago, *Banking System Risk: Charting a New Course.* Chicago: 1989, pp. 515–546.

Bernanke, Ben S., "Nonmonetary Effects of the Financial Crisis in the Propagation of the Great Depression," *American Economic Review,* 73 (June 1983), pp. 257–276.

Bernheim, B. Douglass, "The Crisis in Deposit Insurance: Issues and Options," in Association of Reserve City Bankers, *Capital Issues in Banking.* Washington, D.C.: 1988, pp. 160–197.

Bloch, Ernest, "The Federal Home Loan Bank System," in George F. Break, et al., *Federal Agencies.* Englewood Cliffs, N.J.: Prentice-Hall, 1963, pp. 165–257.

Bloch, Ernest, "Multiple .Regulators: Their Constituencies and Policies," in Yakov Amihud, Thomas S.Y. Ho, and Robert A. Schwartz, eds., *Market Making and the Changing Structure of the Securities Industry.* Lexington, Mass.: D.C. Health, 1985, pp. 155–181.

Bosworth, Barry P., Andrew S. Carron, and Elizabeth H. Rhyne, *The Economics of Federal Credit Programs,* Washington, D.C.: Brookings, 1987.

Boyd, John M., and Arthur J. Rolnick, "A Case for Reforming Federal Deposit Insurance," in Federal Reserve Bank of Minneapolis, *Annual Report, 1988.*

Brewer, Elijah, III, et al., "The Depository Institutions Deregulation and Monetary Control Act of 1980," *Economic Perspectives,* Federal Reserve Bank of Chicago, 4 (September/October 1980), pp. 3–23.

Brewer, Elijah, III, "Full-Blown Crisis, Half-Measure Cure," *Economic Perspectives,* Federal Reserve Bank of Chicago, 13 (November/December 1989a), pp. 2–16.

Brewer, Elijah, III, "S & Ls: Running on Empty," *Chicago Fed Letters,* no. 28, December 1989b.

Brown, Richard A., Joseph A. McKenzie, and Rebel A. Cole, "Going Beyond Traditional Mortgages: The Portfolio Performance of Thrifts," FIS no. 1–90, Financial Industry Studies, Federal Reserve Bank of Dallas., February 1990.

Brumbaugh, R. Dan, Jr., *Thrifts under Siege: Restoring Order to American Banking.* Cambridge, Mass.: Ballinger, 1988.

Brumbaugh, R. Dan, Jr., and Andrew S. Carron, "Thrift Industry Crisis: Causes and Solutions," *Brookings Papers on Economic Activity* (1987, no. 2), pp. 349–377.

Brumbaugh, R. Dan, Jr., Andrew S. Carron, and Robert E. Litan, "Cleaning Up the Depository Institutions Mess," *Brookings Papers on Economic Activity* (1989, no. 1), pp. 243–283.

Brumbaugh, R. Dan, Jr., and Eric I. Hemel, "Federal Deposit Insurance as a Call Option: Implications for Depository Institutions," Research Working Paper no. 116, Office of Policy and Economic Research, Federal Home Loan Bank Board, October 1984.

Brumbaugh, R. Dan, Jr., and Robert E. Litan, "The S&L Crisis: How to Get Out and Stay Out," *Brookings Review,* 7 (Spring 1989), pp. 3–13.

Brumbaugh, R. Dan, Jr., and Robert E. Litan, "The Banks are Worse Off Than You Think," *Challenge,* 33 (January/February 1990), 33 pp. 4–12.

Buser, Stephen A., Andrew W. Chen, and Edward J. Kane, "Federal Deposit Insurance, Regulatory Policy, and Optimal Bank Capital," *Journal of Finance,* 36 (March 1981), pp. 41–60.

Calomiris, Charles W., "Deposit Insurance: Lessons from the Record," *Economic Perspectives,* Federal Reserve Bank of Chicago, 13 (May/June 1989a), pp. 10–30.

Calomiris, Charles W., "Success and Failure in Pre-Depression Bank Liability Insurance," in Federal Reserve Bank of Chicago, *Banking System Risk: Charting a New Course.* Chicago: 1989b, pp. 198–220.

Campbell, Tim S., *Money and Capital Markets.* Glenview, Ill.: Scott, Foresman, 1988.

Carns, Frederick S., "Should the $100,000 Deposit Insurance Limit Be Changed?" *FDIC Banking Review,* 2 (Spring/Summer 1989), pp. 11–19.

Carron, Andrew S., *The Plight of the Thrift Institutions.* Washington, D.C.: Brookings Institution, 1982.

Carron, Andrew S., and R. Dan Brumbaugh, Jr., "The Future of Thrifts in the Mortgage Market," in Federal Reserve Bank of Chicago, *Banking System Risk: Charting a New Course.* Chicago: 1989, pp. 385–400.

Chan, Yuk-Shee, Stuart I. Greenbaum, and Anjan V. Thakor, "Is Fairly Priced Deposit Insurance Possible?" Working Paper no. 152, Banking Reserve Center, Kellogg Graduate School of Management, Northwestern University, October 1988.

Cole, Rebel A., "Insolvency versus Closure: Why the Regulatory Delay in Closing Troubled Thrifts?" mimeo, Financial Industries Studies Department, Federal Reserve Bank of Dallas, February 6, 1990a.

Cole, Rebel A., "Agency Conflicts and Thrift Resolution Costs," mimeo, Financial Industries Studies Department, Federal Reserve Bank of Dallas, February 7, 1990b.

Cole, Rebel A., Robert A. Eisenbeis, and Joseph A. McKenzie, "Excess Returns and Sources of Value in FSLIC-Assisted Acquisitions of Troubled Thrifts," FIS no. 1–89, Financial Industry Studies, Federal Reserve Bank of Dallas, December 1989.

Cole, Rebel A., and Joseph A. McKenzie, "Efficient Diversification of Thrift Institution Portfolios," mimeo, Federal Reserve Bank of Dallas, May 1990.

Cook, Douglas O., and Lewis J. Spellman, "Market Cynicism of Government Guarantees: The Waning Days of the FSLIC," mimeo, University of Texas, January 1990.

Diamond, Douglas W., and Philip H. Dybvig, "Bank Runs, Deposit Insurance, and Liquidity," *Journal of Political Economy,* 91 (June 1983), pp. 401–419.

Dochow, Daniel W., "Letter to Craig A. Simmons, General Accounting Office," Office of Regulatory Activities, Federal Home Loan Bank System, May 8, 1989.

Edwards, Linda N., and Franklin R. Edwards, "Measuring the Effectiveness of Regulation: The Case of Bank Entry Regulation," *Journal of Law & Economics*, 17 (October 1974), pp. 445–460.

Eichler, Ned, *The Thrift Debacle*. Berkeley: University of California Press, 1989.

Eisenbeis, Robert A., "Bank Holding Companies and Public Policy," in George J. Benston, ed., *Financial Services: Changing Institutions and Public Policy*. Englewood Cliffs, N.J.: Prentice Hall, 1983, pp. 127–155.

Eisenbeis, Robert A., "Restructuring Banking," *Challenge*, 33 (January/February 1990), pp. 18–21.

Engelke, George L., "Mark-to-Market Accounting: What *Is* Real Market Value?" *Financial Managers' Statement*, 12 (January/February 1990), pp. 33–34.

England, Catherine, "A Market Approach to the Savings and Loan Crisis," in Edward H. Crane and David Boaz, eds., *An American Vision: Policies for the '90's*. Washington, D.C.: Cato Institute, 1989, pp.

Fabritius, M. Manfred, and William Borges, *Saving the Savings and Loan: The U.S. Thrift Industry and the Texas Experience, 1950–1988*. New York: Praeger, 1989.

Federal Deposit Insurance Corporation, *Deposit Insurance for the Nineties: Meeting the Challenge*. Washington, D.C.: January 4, 1989.

Federal Home Loan Bank Board, *Annual Report, 1982*.

Federal Home Loan Bank Board, *Agenda for Reform, A Report on Deposit Insurance to the Congress*. Washington, D.C.: March 1983a.

Federal Home Loan Bank Board, "Report of the Expanded Task Force on Current Value Accounting," mimeo, Washington, D.C., April 12, 1983b.

Federal Home Loan Bank Board, *Annual Report, 1983c*.

Federal Home Loan Bank Board, *Annual Report, 1984*.

Federal Home Loan Bank Board, *Annual Report, 1985*.

Federal Home Loan Bank Board, *Annual Report, 1987*.

Fischel, Daniel K., Andrew M. Rosenfeld, and Robert S. Stillman, "The Regulation of Banks and Bank Holding Companies," *Virginia Law Review*, 73 (March 1987), pp. 301–338.

Flannery, Mark J., "Deposit Insurance Creates a Need for Bank Regulation," *Business Review*, Federal Reserve Bank of Philadelphia (January/February 1982), pp. 17–27

Flannery, Mark J., "An Economic Evaluation of Bank Securities Activities before 1933," in Ingo Walter, ed., *Deregulating Wall Street: Commercial Bank Penetration of the Corporate Securities Market*. New York: John Wiley & Sons, 1985, pp. 67–87.

Flannery, Mark J., "Pricing Deposit Insurance When the Insurer Measures Risk with Error," in Federal Reserve Bank of Chicago, *Banking System Risk: Charting a New Course*. Chicago: 1989, pp. 70–100.

Friend, Irwin, ed., *Study of the Savings and Loan Industry,* Volumes I–IV. Washington, D.C.: U.S. Government Printing Office, July 1969.

Garcia, Gillian, et al., "The Garn–St Germain Depository Institutions Act of 1982," *Economic Perspectives,* Federal Reserve Bank of Chicago, 6 (April 1983), pp. 3–31.

Golbe, Devra L., "The Effects of Imminent Bankruptcy on Stockholder Risk Preferences and Behavior," *Bell Journal of Economics,* 12 (Spring 1981), pp. 321–328.

Goodman, Laurie S., and Sherrill Shaffer, "The Economics of Deposit Insurance: A Critical Evaluation of Proposed Reform," *Yale Journal on Regulation,* 2 (1984), pp. 145–162.

Gorton, Gary, "Public Policy and the Evolution of Banking Markets," in Federal Reserve Bank of Chicago, *Banking System Risk: Charting a New Course.* Chicago: 1989, pp. 233–252.

Guttentag, Jack M., and Richard J. Herring, "Restructuring Depository Institutions," mimeo, Wharton School, University of Pennsylvania, 1988.

Hempel, George H., Donald G. Simonson, Marvin L. Carlson, Marsha Simonson, and Marcia M. Cornett, "Market Value Accounting for Financial Service Companies," mimeo, National Center for Financial Service Companies, University of California, Berkeley, 1989.

Hirschhorn, Eric, "Depositor Risk Perceptions and the Insolvency of the FSLIC," mimeo, Office of Thrift Supervision, December 1989.

Hirschhorn, Eric, "Interest Rates on Thrift Certificates of Deposit," Special Report, Office of the Chief Economist, Office of Thrift Supervision, February 1990.

Horvitz, Paul M., "The Case Against Risk-Related Deposit Insurance Premiums," *Housing Finance Review,* 2 (July 1983a), pp. 253–263.

Horvitz, Paul M., "Market Discipline is Best Provided by Subordinated Creditors," *American Banker* (July 15, 1983b), p. 4.

Horvitz, Paul M., "Deposit Insurance after Deregulation: A Residual Role for Regulation," in Federal Home Loan Bank of San Francisco, *Identification and Control of Risk in the Thrift Industry.* San Francisco: 1983c, pp. 67–94.

Horvitz, Paul M., "Subordinated Debt is Key to New Bank Capital Requirements," *American Banker* (December 31, 1984), p. 5.

Horvitz, Paul M., "The FSLIC Crisis and the Southwest Plan," *American Economic Review,* 79 (May 1989a), pp. 146–150.

Horvitz, Paul M., "Implications of the Texas Experience for Financial Regulation," in Federal Reserve Bank of Chicago, *Banking System Risk: Charting a New Course.* Chicago: 1989b, pp. 301–311.

Jackson, Brooks, *Honest Graft: Big Money and the American Political Process.* New York: Alfred A. Knopf, 1988.

Jacobs, Donald P., and Almarin Phillips, "Overview of the Commission's Philosophy and Recommendations," in Federal Reserve Bank of Boston, *Policies for a More Competitive Financial System.* Boston: 1972.

Jensen, Michael C., and William H. Meckling, "Theory of the Firm: Managerial Behavior, Agency Costs, and Ownership Structure," *Journal of Financial Economics,* 3 (October 1976), pp. 305–360.

Kane, Edward J., "Short-Changing the Small Saver: Federal Discrimination Against the Small Saver during the Vietnam War," *Journal of Money, Credit and Banking,* 2 (November 1970), pp. 513–522.

Kane, Edward J., "Deregulation, Savings and Loan Diversifications and the Flow of Housing Finance," in Federal Home Loan Bank of San Francisco, *Savings and Loan Asset Management under Deregulation.* San Francisco: 1980, pp. 81–109.

Kane, Edward J., "S&Ls and Interest-Rate Regulation: The FSLIC as an In-Place Bailout Program," *Housing Finance Review,* 1 (July 1982a), pp. 219–243.

Kane, Edward J., "Metamorphosis in Financial Services Delivery and Production," in Federal Home Loan Bank of San Francisco, *Strategic Planning for Economic and Technological Change in the Financial Services Industry.* San Francisco: 1982b, pp. 49–66.

Kane, Edward J., "The Role of Government in the Thrift Industry's Net Worth Crisis," in George J. Benston, ed., *Financial Services: The Changing Institutions and Government Policy.* Englewood Cliffs, N.J.: Prentice-Hall, 1983, pp. 156–184.

Kane, Edward J., *The Gathering Crisis in Federal Deposit Insurance.* Cambridge, Mass.: MIT Press, 1985.

Kane, Edward J., "Appearance and Reality in Deposit Insurance: The Case for Reform," *Journal of Banking and Finance,* 10 (June 1986), pp. 175–188.

Kane, Edward J., "How Incentive-Incompatible Deposit-Insurance Funds Fail," Prochnow Report no. PR–014, Prochnow Educational Foundation, Madison, Wisc., 1988.

Kane, Edward J., *The S&L Insurance Mess: How Did It Happen?* Washington, D.C.: Urban Institute Press, 1989a.

Kane, Edward J., "The Bush Plan is No Cure for the S&L Malady," *Challenge,* 36 (November/December 1989b), pp. 39–43.

Kaplan, Daniel P., "The Changing Airline Industry," in Leonard W. Weiss and Michael W. Klass, eds., *Regulatory Reform: What Actually Happened.* Boston: Little, Brown, 1986, pp. 40–77.

Karaken, John H., "Deposit Insurance Reform or Deregulation is the Cart Not the Horse," *Quarterly Review,* Federal Reserve Bank of Minneapolis, 7 (Spring, 1983), pp. 1–9.

Karaken, John H., and Neil Wallace, "Deposit Insurance and Bank Regulation," *Journal of Business,* 51 (July 1978), pp. 413–438.

Kaufman, George G., "The Truth about Bank Runs," in Catherine England and Thomas Huertas, eds., *The Financial Services Revolution: Policy Directions for the Future.* Boston: Kluwer, 1988, pp. 9–40.

Keeler, Theodore E., "The Revolution in Airline Regulation," in Leonard W. Weiss and Michael W. Klass, eds., *Case Studies in Regulation: Revolution and Reform.* Boston: Little, Brown, 1981, pp. 53–85.

Keeley, Michael C., "Deposit Insurance, Risk, and Market Power in Banking," in Federal Reserve Bank of Chicago, *Banking System Risk: Charting a New Course.* Chicago: 1989, pp. 101–116.

Kendall, Leon T., *The Savings and Loan Business.* Englewood Cliffs, N.J.: Prentice-Hall, 1962.

Klinkerman, Steve, "Profits in Texas S&L Deals Only Average," *American Banker* (March 15, 1990), p. 19.

Kormendi, Roger C., Victor L. Bernard, S. Craig Pirrong, and Edward A. Snyder, *Crisis Resolution in the Thrift Industry,* Report of the Mid America Institute Task Force on the Thrift Crisis, February 1989.

Litan, Robert E., *What Should Banks Do?* Washington, D. C.: Brookings Institution, 1987.

MacDonald, James M., "Railroad Deregulation, Innovation and Competition: Effects of the Staggers Act on Grain Transportation," *Journal of Law & Economics,* 32 (April 1989), pp. 63–95.

Mahoney, Patrick I., and Alice P. White, "The Thrift Industry in Transition," *Federal Reserve Bulletin,* 71 (March 1985), pp. 137–156.

Marcus, Alan J., "Deregulation and Bank Financial Policy," *Journal of Banking and Finance,* 8 (December 1984), pp. 557–565.

Marvell, Thomas B., *The Federal Home Loan Bank Board.* New York, Praeger, 1969.

McCall, Alan S., and John T. Lane, "Multi-Office Banking and the Safety and Soundness of Commercial Banks," *Journal of Bank Research,* 11 (Summer 1980), pp. 87–94.

Merton, Robert C., "An Analytic Derivation of the Cost of Deposit Insurance and Loan Guarantees," *Journal of Banking and Finance,* 1 (June 1977), pp. 3–11.

Merton, Robert C., "On the Cost of Deposit Insurance When There are Surveillance Costs," *Journal of Business,* 51 (July 1978), pp. 439–452.

Meyer, John R., and Clinton V. Oster, eds., *Airline Deregulation: The Early Experience.* Boston: Auburn House, 1981.

Mills, Edwin S., and Lawrence J. White, "Government Policies Toward Automotive Emissions Control," in Ann F. Friedlander, ed., *Approaches to Controlling Air Pollution.* Cambridge, Mass.: MIT Press, 1978, pp. 348–309.

Moore, Thomas G., "Rail and Trucking Deregulation," in Leonard W. Weiss and Michael W. Klass, eds., *Regulatory Reform: What Actually Happened.* Boston: Little, Brown, 1986, pp. 14–39.

Morrison, Stephen, and Clifford Winston, *The Economic Effects of Airline Deregulation.* Washington, D.C.: Brookings Institution, 1986.

Murton, Arthur J., "Bank Intermediation, Bank Runs, and Deposit Insurance," *FDIC Banking Review,* 2 (Spring/Summer 1989), pp. 1–10.

National Association of Mutual Savings Banks, *Mutual Savings Banking: Basic Characteristics and Role in the National Economy.* Englewood Cliffs, N.J.: Prentice-Hall, 1962.

Ogden, William, Jr., Nanda Rangan, and Thomas Stanley, "Risk Reduction in S&L Mortgage Loan Portfolios through Geographic Diversification," *Journal of Financial Services Research,* 2 (February 1989), pp. 39–48.

Peltzman, Sam, "Entry in Commercial Banking," *Journal of Law & Economics,* 8 (October 1965), pp. 11–50.

Peltzman, Sam, "Toward a More General Theory of Regulation," *Journal of Law & Economics,* 19 (August 1976), pp. 211–240.

Phelan, Richard J., *Record of the Special Outside Counsel in the Matter of Speaker James C. Wright, Jr.,* Committee on Standards of Official Conduct, U.S. House of Representatives. Washington, D.C.: U.S. Government Printing Office, February 21, 1989.

Phillips, Almarin, and Donald P. Jacobs, "Reflections on the Hunt Commission," in George J. Benston, ed., *Financial Services; The Changing Institutions and Government Policy.* Englewood Cliffs, N.J.: Prentice-Hall, 1983, pp. 235–265.

Pilzer, Paul Z., *Other People's Money: The Inside Story of the S&L Mess.* New York: Simon and Schuster, 1989.

Pizzo, Stephen, Mary Fricker, and Paul Muolo, *Inside Job: The Looting of America's Savings and Loans.* New York: McGraw-Hill, 1989.

Posner, Richard A., "Theories of Economic Regulation," *Bell Journal of Economics and Management Science,* 5 (Autumn 1974), pp. 335–358.

Pusey, Allen, "Fast Money and Fraud," *New York Times Magazine* (April 23, 1989), pp. 30–35.

Rolnick, Arthur J., and Warren E. Weber, "New Evidence on the Free Banking Era," *American Economic Review,* 73 (December 1983), pp. 1080–1091.

Scott, Kenneth E., "Deposit Insurance and Bank Regulation: The Policy Choices," *Business Lawyer,* 44 (May 1989), pp. 907–933.

Scott, Kenneth E., and Thomas Mayer, "Risk and Regulation in Banking: Some Proposals for Deposit Insurance," *Stanford Law Review,* 23 (May 1971), pp. 857–902.

Simonson, Donald G., and George H. Hempel, "Historical Perspective in Accounting for Financial Institutions' Performance," in Federal Reserve Bank of Chicago, *Banking System Risk: Charting a New Course.* Chicago: 1989, pp. 504–514.

Smith, Fred, and Melanie S. Tammen, "Plugging America's Financial Black Holes: Reforming the Federal Deposit Insurance System," *Consumer Finance Law Quarterly Report,* 43 (Winter 1989), pp. 44–48.

Spellman, Lewis J., *The Depository Firm and Industry: Theory, History, and Regulation.* New York: Academic Press, 1982.

Sprague, Irvine H., *Bailout: An Insider's Account of Bank Failures and Rescues.* New York: Basic Books, 1986.

Stigler, George J., "The Theory of Regulation," *Bell Journal of Economics and Management Science,* 2 (Spring 1971), pp. 3–21.

Stoll, Hans R., *Regulation of Securities Markets: An Examination of the the Effects of Increased Competition.* Monograph Series in Finance and Economics no. 1979–2, Solomon Brothers Center for the Study of Financial Institution, Graduate School of Business Administration, New York University, 1979.

Stoll, Hans R., "Revolution in the Regulation of Securities Markets: An Examination of the Effects of Increased Competition," in Leonard W. Weiss and Michael W. Klass, eds., *Case Studies in Regulation: Revolution and Reform.* Boston: Little, Brown, 1981, pp. 12–52.

Strunk, Norman, and Fred Case, *Where Deregulation Went Wrong: A Look at the Causes Behind Savings and Loan Failures in the 1980s.* Chicago: U.S. League of Savings Institutions, 1988.

Tammen, Melanie S., "The Savings & Loan Crisis: Which Train Derailed— Deregulation or Deposit Insurance?" *Journal of Law & Politics,* 6 (Winter 1990), pp. 311–342.

Tinic, Seha M., and Richard R. West, "The Securities Industry Under Negotiated Brokerage Commissions: Changes in the Structure and Performance of the New York Stock Exchange Member Firms," *Bell Journal of Economics,* 11 (Spring 1980), pp. 29–41.

Tuccillo, John A., *Housing Finance: A Changing System in the Reagan Era.* Washington, D.C.: Urban Institute Press, 1983.

U.S. Congress, House of Representatives, Committee on Government Operations, *Combating Fraud, Abuse, and Misconduct in the Nation's Financial Institutions: Current Federal Efforts are Inadequate.* Seventy-Second Report. Washington, D.C.: U.S. Government Printing Office, October 13, 1988.

U. S. Department of the Treasury, *Geographic Restrictions on Commercial Banking in the United States: The Report of the President.* Washington, D.C., 1981.

Walter, Ingo, ed., *Deregulating Wall Street: Commercial Bank Penetration of the Corporate Securities Market.* New York: John Wiley & Sons, 1985.

Weiss, Leonard W., and Michael W. Klass, eds., *Case Studies in Regulation: Revolution and Reform.* Boston: Little, Brown, 1981.

Weiss, Leonard W., and Michael W. Klass, eds., *Regulatory Reform: What Actually Happened.* Boston: Little, Brown, 1986.

White, Eugene N., *The Regulation and Reform of the American Banking System, 1900– 1929.* Princeton, N.J.: Princeton University Press, 1983.

White, Eugene N., "Commercial Banks and Securities Markets: Lessons of the 1920s and 1930s for the 1980s and 1990s," in Federal Reserve Bank of Chicago, *Banking System Risk: Charting a New Course.* Chicago: 1989, pp. 258–266.

White, Lawrence J., "Price Regulation and Quality Rivalry in a Profit Maximizing Model: The Case of Bank Branching," *Journal of Money, Credit and Banking,* 7 (February 1976), pp. 97–106.

White, Lawrence J., *Reforming Regulation: Processes and Problems.* Englewood Cliffs, N.J., Prentice-Hall, 1981.

White, Lawrence J., *The Regulation of Air Pollutant Emissions from Motor Vehicles.* Washington, D.C.: American Enterprise Institute, 1982.

White, Lawrence J., "The Partial Deregulation of Banks and Other Depository Institutions," in Leonard W. Weiss and Michael W. Klass, eds., *Regulatory Reform: What Actually Happened.* Boston: Little, Brown, 1986, pp. 169–209.

White, Lawrence J., "Mark to Market is Vital to FSLIC and Valuable to Thrifts," *Outlook of the Federal Home Loan Bank System,* 4 (January/February 1988a), pp. 20–24.

White, Lawrence J., "Litan's *What Should Banks Do?* A Review Essay," *Rand Journal of Economics,* 19 (Summer 1988b), pp. 305–315.

White, Lawrence J., "Market Value Accounting: An Important Part of the Reform of the Deposit Insurance Systems," in Association of Reserve City Bankers, *Capital Issues in Banking.* Washington, D.C.: 1988c, pp. 226–242.

White, Lawrence J., "The Reform of Federal Deposit Insurance," *Journal of Economic Perspectives,* 3 (Fall 1989a), pp. 11–29.

White, Lawrence J., "Marking to Market Recognizes Realities," *American Banker* (December 12, 1989b), p. 6.

White, Lawrence J., "Mark-to-Market: A (Not So) Modest Proposal," *Financial Managers' Statement,* 12 (January/February 1990a), pp. 27–37.

White, Lawrence J., "The Case for Mark-to-Market Accounting," *Secondary Mortgage Markets,* 7 (Summer 1990b), pp. 2–4.

White, Lawrence J., "The Value of Market Value Accounting for the Deposit Insurance System," *Journal of Accounting, Auditing, and Finance,* 6 (April 1991).

White, Lawrence J., and Edward L. Golding, "Collateralized Borrowing at Thrifts Poses Risk to FSLIC," *American Banker* (February 24, 1989), p. 4.

Woerheide, Walter J., *The Savings and Loan Industry: Current Problems and Possible Solutions.* Westport, Conn.: Quorum Books, 1984.

Woodward, G. Thomas, "FSLIC, the Budget, and the Economy," mimeo, Congressional Research Service, Library of Congress, January 12, 1989.

Index